CONTENTS

LIST OF FIGURES

Mersey Estuary MANAGEMENT PLAN

A Strategic Policy Framework

The University of Liverpool Study Team

Mersey Basin Campaign

ISBN 0-85323-870-7

Liverpool University Press

1995

Liverpool University Press

Senate House, Abercromby Square, Liverpool, L69 3BX

Mersey Basin Campaign

Designed by Cheshire County Council, **Central Graphic Design** 01244 603112. J8718.

Maps produced by the **Government Office for the North West.**

FOREWORD

The Mersey Estuary Management Plan presented here marks the culmination of more than three years' intensive effort on the part of the Mersey Basin Campaign's Estuary Project Group and their consultants, the University of Liverpool Study Team.

It is undoubtedly a pioneering venture. The Mersey is the first highly-developed estuary in Western Europe for which a management plan has been prepared. Estuaries of this kind present a particularly difficult challenge in planning terms, given their inherent complexity, the wide range of issues to be confronted, and the large number of organisations with a vested interest in river activities.

Because there were no 'models' to follow, it has been necessary to develop a completely new approach. The Plan's strategic policy framework establishes for the first time a common base for the systematic development of estuary policies and management measures. Individual policy areas have also received novel treatment: particular examples are to be found in the sections dealing with estuary resources and economic development.

The Plan is innovative in another important respect. This concerns the close working relationship that has developed between the University Study Team and the Estuary Project Group. During the three-year commission, the Team has played a central role in all aspects of plan preparation, including public consultation exercises and the organisation of annual Mersey Estuary Conferences, as well as the technical work associated with the drafting of the Plan. Very much in the spirit of the Campaign, the University Study Team has sought to develop shared ownership of the Plan by involving as many interested parties as possible

at all stages in the commission. The Estuary Project Group's Technical Steering Group has provided effective guidance throughout, making a significant contribution to the topic reports and to the production of the final document. The University's role has been crucial: in terms of its independent stance on Estuary matters, its wide range of technical expertise, and its firm commitment to the local region.

Publication of the Plan brings with it a major responsibility: that of ensuring that the document continues to influence Estuary policy and decision-makers. For it to be successful, the Plan must provide the focus of an on-going management process, involving all of those who are in a position to take action on the Estuary both now and in years to come. My view is that it presents us with an outstanding opportunity.

I would like to offer my congratulations to all who have been involved in the production of the Plan. Its publication comes at an important time in the history of the Mersey and its tributaries. All the signs are that the tide of pollution is turning at last and the previous perspective for the Mersey as the most polluted river system in Europe has changed to a recognition that it is now the most improved.

Brian Alexander
Chairman, Mersey Basin Campaign.

Mersey Basin Campaign

THE UNIVERSITY OF LIVERPOOL STUDY TEAM

Professor Peter Batey (Director)

Lever Professor of Town and Regional Planning and Head of the
Department of Civic Design

Miss Anne Dennier

Senior Fellow, Department of Civic Design

Ms. Sue Kidd

Lecturer, Department of Civic Design

Dr. David Massey

Lecturer, Department of Civic Design

Dr. Geoff Woodcock

Lecturer, School of Politics and Communication Studies

Postgraduate Student Research Assistants

Guy Currey

Helen Curry

Anne Dugdale

Joanne Farrar

Jonathan Francis

Karen Hearnshaw

Emma Heron

Esther Keen

Andrea McGuinness

Debbie Moore

Henry Oliver

Louise Shore

Mark Stevenson

Edward Taylor

Rosemary Vernon

Andrea Voyce

Hamish Walke

Emma Williamson

Liz Wood-Griffiths

MERSEY BASIN CAMPAIGN ESTUARY PROJECT GROUP

Chairman

1993 -95

Tim Cox

Sefton Metropolitan Borough Council

1991 - 93

Michael Hayes

Liverpool City Council

Vice Chairman

1991 - 95

Tom Baty

Manchester Ship Canal Company

Secretary

1993 - 95

Ceri Jones

Sefton Metropolitan Borough Council

1991 - 93

Ian Wray

Liverpool City Council

Chairman, Technical Steering Group, Estuary Management Plan

1992 - 95

Stuart Roberts

Cheshire County Council

INTRODUCTION

The Mersey Estuary is of great importance to the economy and environment of the region. It is a major trade route and the location for significant industrial installations and urban centres. It is also a powerful natural system with high nature conservation interest.

The River Mersey drains a catchment of some 2000 square miles containing the urban concentrations of most of Greater Manchester and Merseyside. As a consequence, the Estuary has received very heavy pollution loads. Extensive dockland and industrial development along its banks, coupled with rapid growth of urban areas all draining into the Estuary, led to an unenviable reputation as the worst polluted estuary in Europe.

Over the past twenty years, however, massive investment in water quality improvements, reclamation of derelict land and new riverside development has created conditions in which it is possible to think of the Estuary as an opportunity rather than a problem, and to give very careful thought to its long-term future.

A Management Plan for the Estuary

Improving the waterside environment has been a major aim of the Mersey Basin Campaign since its inception in 1985, as a partnership between central and local government, private and voluntary sector bodies. The Campaign covers the whole Mersey catchment and is divided into five Project Groups. The Estuary Project Group comprises some forty organisations including the bankside local authorities, major companies with a direct role in the Estuary, as well as statutory agencies with responsibility for different aspects of the Estuary.

The Group felt that a Management Plan for the Estuary was needed and commissioned the Department of Civic Design at the University of Liverpool to prepare a plan and work began in February 1992. Half the cost has been met by the Department of the Environment's Mersey Basin Campaign Unit with the remainder being met by the National Rivers Authority, English Nature and the local authorities of Cheshire County, Ellesmere Port and Neston, Halton, Liverpool, Sefton, Vale Royal, Warrington and Wirral.

The Management Plan for the Mersey Estuary which is presented here is the culmination of three years' work by the University Study Team, aided throughout by a Technical Steering Group. This small group of individuals, selected for their special knowledge of particular aspects or areas of the Estuary, provided detailed advice and guidance on behalf of the larger Estuary Project Group. More information about the commission, and the University's role in preparing the Plan, is provided in Appendix 3.

The Plan is founded on up-to-date research and careful attention to the views of organisations which have a direct role or interest in the current management of the Estuary. It sets out the main issues that need to be tackled and the range of policies and measures required to provide a framework for future management.

Two factors have been central to the commissioning of work on the Management Plan. First, there has been growing recognition of the significance of the Estuary as an environmental resource of international, national and regional importance. Secondly, there has been a developing awareness of the existing and potential links between economic development and the Estuary. On the negative side these links include the scope for development to damage the estuarine environment as witnessed by the current state of water quality in the Mersey. On the positive side, the Estuary is seen not only as critical to the economic well-being of the area's maritime economy, but also as a natural asset which could provide a springboard for new economic development, as well as recreation and tourist facilities.

Helping to achieve an appropriate balance between development needs and aspirations and the protection and enhancement of Estuary resources for future generations can be seen as the fundamental role of the Estuary Management Plan. Reconciling these two aspirations is at the heart of sustainable development.

Current developments reinforce this point. The Management Plan is being prepared at a time when the Estuary seems to be emerging from a 'dark age' of pollution and neglect and when its economic strengths appear to be reviving after a period of decline. A recent report produced by the National Rivers Authority (*The Mersey Estuary: A Report on Environmental Quality, 1995*) draws attention to some of the achievements to date:

- Over £1 billion has been invested in the Mersey Basin Clean-up.
- The pollution load in rivers flowing into the Estuary has been reduced by 80% over the last twenty five years.
- Mercury discharges have been reduced by more than 90% in the last fifteen years.
- Radiometric dating of marsh sediments has shown reductions in heavy metal concentrations, with some possibly approaching pre-industrial levels.
- The Estuary has become a haven for wildfowl and waders with numbers rising by more than 60% over the last decade. Increasing numbers of fish are now returning to the Estuary, with more than 35 species having been found in recent years.

Such evidence clearly points to the fact that the biological health of the Estuary has improved dramatically over the past twenty or so years; in addition there have been significant improvements in the bankside environment; there is a far greater understanding of the Estuary as a natural system; and there is a preparedness to work together in solving problems.

Nevertheless, the scale of the issues identified in the Management Plan must not be underestimated. The Mersey is, after all, one of Britain's largest estuaries. It has one of the highest tidal ranges in the world and there are intractable natural forces to be dealt with. It is a refuge for wildlife, but at the same time it is at the centre of an intensively developed urban region. The land uses ranged around its banks present a complex picture of economic and development issues.

The Study Brief

The commissioning brief, issued in February 1992, made it clear that the Management Plan is to provide an advisory framework for the future management of the Estuary. This framework is intended to enable existing interests to be safeguarded; new development proposals to be evaluated; and the full potential of the Estuary as a natural resource to be realised.

The brief identified five specific aims for the Plan:
- to focus attention on the Estuary as one of the Mersey region's most important environmental assets and convey a positive image of the area as a unique conurbation with an enormous water resource (with recreational and tourist potential) at its core;
- to provide the basis for an agreed and coordinated programme of environmental action and creative conservation to be implemented by

View from Hightown looking up the Outer Estuary towards Liverpool and the 'Narrows'

the commissioning partners and others;

• to set out proposals for the management of river-based recreation and for the protection of ecological assets;

• to establish part of the technical basis to enable the local authorities and others to respond to major development initiatives on the Estuary;

• to enable the commissioning partners to speak with an informed and authoritative voice on matters affecting the Estuary.

Study Area : The Estuary Zone

One of the first tasks of the University Team was to define the Study Area. It was recognised that many aspects of work would require a very broad perspective (for example, nature conservation issues have an international dimension and water quality issues require a regional view), but the Team felt that the definition of a core area would be valuable in providing a general focus for the Study.

Factors influencing the final definition of the core area - referred to in the Plan as the Estuary Zone and illustrated on the key map on page 104 included:

• the tidal limit of the Estuary;

• the extent of the 10m contour;

• an assessment of the existing and potential landward influences of the Estuary and identification of clear inland boundaries;

• the boundaries of neighbouring management plan exercises for the Dee and Ribble Estuaries;

• the coverage of dredging deposit grounds in Liverpool Bay;

• the jurisdiction of maritime authorities.

Two features of the boundary are worthy of comment. The first concerns the North Wirral Foreshore which has been included in both the Mersey and the Dee Management Plan areas. It was felt to be unhelpful to allocate this stretch of coast exclusively to one area, given the particular issues involved which have a significant bearing on the policies adopted in both plans. The second feature to note is that the Estuary Zone includes territory in two of the five project group areas of the Mersey Basin Campaign: the Estuary Project Group area and the Southern Catchment Group area. Potential problems of coordination have been avoided by ensuring that all affected parties are represented on the Estuary Project Group, the body responsible for the Estuary Management Plan.

The Basic Approach

The issues to be addressed in the Plan were identified in two main ways: by means of topic reports in which up-to-date information on relevant subjects was assembled and analysed; and as the result of consultation exercises involving statutory agencies, voluntary groups and the private sector.

A total of fifteen topic reports have been prepared on subjects ranging from Water Quality and Nature Conservation to Emergency Planning. These reports represent one of the major outputs of the University Study Team's work and constitute a substantial information resource. A full list of the documents that have been produced is contained in Appendix 4.

Consultation has been undertaken in a number of ways. In the first year of the commission, a questionnaire survey was carried out among voluntary groups and private sector organisations. Statutory agencies were also surveyed and in many cases interviewed. In this way not only were strategic issues identified, but also a clearer understanding was obtained

The Inner Estuary at Ellesmere Port

of the powers and functions of the various bodies involved in the operation and management of the Estuary, including the interests of local businesses.

In the second year, Area Workshops were held in Bootle, Chester and Warrington at which Area Issue Reports were discussed. These workshops provided an opportunity for the Study Team to obtain input and feedback from councillors, officials and voluntary group members with detailed local knowledge of particular sections of the Estuary. In addition, a study day was organised, under the auspices of the Mersey Basin Trust, for the benefit of voluntary sector groups.

The Draft Management Plan was launched in October 1994, part-way through the third year. This was followed by a large number of public meetings held in different parts of the Estuary Zone, at which there was detailed discussion of the Draft Plan's policies and proposals. During a ten week consultation period, more than sixty organisations and individuals provided written comments on the Draft Plan. These were subsequently analysed carefully by the Study Team and in each case a response was framed. A detailed record of the consultation exercise is contained in a separate report (see Appendix 4). The final version of the Plan reflects the changes that have been made as a consequence of the consultation exercise.

Mersey Estuary Conferences, held in March each year since work on the Plan began in 1992, have provided an opportunity for the Study Team to obtain feedback on its work from a wide range of interested parties. In many cases the contacts made at the Conferences have led to further, more informal meetings which have helped to inform and clarify the work on the Plan.

The formal consultation exercise was aimed specifically at obtaining feedback on the policies contained in the Plan. It did not cover proposals for an organisational structure in relation to implementation. These have been the subject of separate consideration by the Estuary Project Group. The proposals were discussed at the Fourth Mersey Estuary Conference in March 1995 and have been modified to take account of comments received. They are included as a separate section in the present document to reflect their different status.

Structure of the Plan

In developing policies which address the strategic issues that have been identified, the Team felt that it was very important to establish a clear structure or framework. This framework would provide the basis for understanding the Estuary's complexity and would serve as a ready means of communicating the Team's proposals. While constituent elements might change over time, the framework would be relatively robust and would represent a stable feature of the Plan.

At the heart of the framework is a Vision Statement which sets out in general terms the main aspirations of the Plan. This Statement, shown in Figure 1, is intended to encapsulate the specific aims of the Plan, as laid down in the commissioning brief. The Vision Statement can itself be broken down into four main subject areas : Estuary Resources, Economic Development, Recreation and Implementation. Associated with these are ten strategic policy areas and strategic objectives.

The Upper Estuary looking towards Runcorn Bridge

Figure 1: Vision Statement

The Mersey Estuary Management Plan will provide a framework for coordinated action. The Plan will be a key instrument in addressing critical management issues so as to secure the sustainable development of the Mersey Estuary and to maintain and develop its position as one of the region's most values environmental assets.

The Management Plan is based on a vision of the future of the Mersey Estuary as one of the cleanest developed estuaries in Europe, where the quality and dynamics of the natural environment are recognised and respected and are matched by a high quality built environment, a vibrant maritime economy, and an impressive portfolio of estuary-related tourism and recreation facilities.

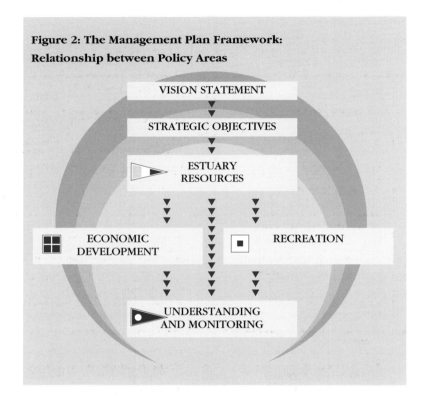

Figure 2: The Management Plan Framework: Relationship between Policy Areas

The relationship between the Vision Statement and the main subject areas is set out in Figure 2. The four main areas (Estuary Resources, Economic Development, Recreation and Implementation) each form the subject of separate chapters in the Plan. One of the four areas, Estuary Resources (which includes Estuary Dynamics, Water Quality and Pollution Control, Biodiversity, and Land Use and Development) has an over-arching function within the Plan. The Estuary Resources section sets a context for the other sections of the Plan and contains a number of key policies which are fundamental to the whole document.

Figure 3 shows how the Strategic Objectives and Policy Areas correspond to the Vision Statement.

Figure 3: The Relationship between the Vision Statement, Policy Areas and Objectives

VISION STATEMENT	ESTUARY RESOURCES		STRATEGIC OBJECTIVE
'The Management Plan is based on a vision of the future of the Estuary as one of the cleanest developed estuaries in Europe, where the quality and dynamics of the natural environment are recognised and respected and are matched by a high quality built environment...	**1**	**Estuary Dynamics**	*To allow the Estuary to function as naturally as possible and in a self-sustaining way by controlling human interference in inter-tidal and marine areas having regard to the natural conditions and processes of the Estuary and Liverpool Bay.*
	2	**Water Quality**	*To support continuing improvements in water, air, land, noise and light quality and the adoption of environmental good practice within the Estuary Zone.*
	3	**Biodiversity**	*To conserve, and where relevant restore, the natural biodiversity of the Estuary Zone.*
	4	**Land Use and Development**	*To promote careful stewardship of land resources, landscape and townscape within the Estuary Zone.*
	ECONOMIC DEVELOPMENT		
'a vibrant maritime economy...	**5**	**Commercial Navigation and Port Development**	*To support the continued commercial and economic development of the Estuary's ports and port-related employment areas compatible with the Management Plan's environmental policies.*
	6	**Urban Regeneration**	*To promote the regeneration of the Estuary Zone through maintaining and realising the distinctive potential of its existing developed waterfront and bankside locations and in adjoining areas.*
and an impressive portfolio of estuary-related tourism,...	**7**	**Tourism**	*To realise the potential of the Estuary as a focus for tourism.*
	RECREATION		
'sport and recreation facilities.	**8**	**Informal and Shore-based Recreation**	*To maintain, enhance and, where appropriate, extend public access to the shores of the Estuary so that people may enjoy informal sport and recreation in safety.*
	9	**Water-based Sport and Recreation**	*To protect existing water-based recreation facilities and promote the appropriate development of new opportunities on the Estuary.*
	IMPLEMENTATION		
The Plan will provide a framework for coordinated action. It will be a key instrument in addressing critical management issues so as to secure the sustainable development of the Estuary and to maintain and develop its position as one of the region's most valued environmental assets.'	**10**	**Understanding and Monitoring**	*Steps should be taken to provide adequate management information to develop understanding and awareness of the natural dynamics of the Estuary and the inter-relation of social and economic activity - including the objectives and policies of the management planning process - with these natural factors.*

The main body of the text in the Plan is devoted to the strategic framework and to the policies which fit within that framework. The ten Policy Areas are specified, each with an accompanying objective. Individual policies are numbered according to the Policy Area to which they refer, so that, for example, NV6 is the sixth policy within the Policy Area referring to Commercial Navigation and Port Development (NV). In cases where specific actions are proposed, they are presented in the form of Management Measures : PC1.4 is, for example, the fourth Management Measure associated with Policy PC1 (Water Quality and Pollution Control). A rationale is provided for each section, Policy Area, Policy and Management Measure, explaining the thinking behind the proposal.

The Plan contains four appendices. The first presents an Index of Management Measures, drawing on measures throughout the Plan and classifying them according to the action that is proposed. In the second an indication is given of how the Management Plan policies may be used in the assessment of major development projects. Three hypothetical case studies are examined as illustrations. In the third appendix background information about the preparation of the Plan is presented : the Plan's origins, the commissioning brief, and the approach that has been adopted by the University Study Team. The fourth appendix lists the documents produced by the Study Team and gives details of how to obtain copies.

Maps included in the Plan

Extensive use has been made of maps to summarise spatially-referenced information in the Plan. Of the four maps that have been included, three are thematic and are linked to the subject material of individual chapters : Estuary Resources; Economic Development; and Recreation. A fourth map, located at the end of the Plan, serves as a key map, and is intended to synthesise the main information contained in the other maps. In all of the maps, the focus of attention is the Estuary Zone.

The Estuary Resources map (Map 1) brings together for the first time comprehensive information on official nature conservation designations within the Estuary. It also describes, in diagrammatic terms, the areas of open coast and areas that are at risk of flooding. The latter information should be interpreted carefully and should not be treated as a substitute for more detailed information that is held by the National Rivers Authority.

The Economic Development map (Map 2) presents information related to port development, urban regeneration and tourism. It defines, again diagrammatically, those areas within the Estuary that are the subject of Permitted (Port) Development Rights. It indicates sites with potential for urban regeneration, and also shows the location of Estuary-related tourist facilities.

Recreation is the theme of Map 3. Using a different approach from that adopted in the other maps, the Recreation map presents an appraisal of the current use and future potential of the main geographical components of the Estuary. This information refers to both shore- and water-based recreation. The map also indicates those locations where conflicts can occur between different kinds of recreation and where certain recreation activities are restricted because of problems of access to the water.

Relationship to Other Plans and Policies

The Management Plan makes frequent reference to other plans and policy documents which also have an important part to play in relation to the Estuary. All of the local authorities in the area have a statutory land use planning function and are required under planning legislation to prepare development plans which set out the land use policy for their areas. The

Unitary Development Plans, District Local Plans and the Cheshire Structure Plan fall into this category. A broader context for such plans is set by Regional Planning Guidance, issued by Central Government, following consultation with the region's local authorities.

Central Government policy has a more direct bearing on the Estuary Management Plan as a result of the policy documents issued by particular departments. Planning Policy Guidance Note 20 (PPG20), on Coastal Planning, was published by the Department of the Environment in 1992 and is arguably the most significant of these. Also important, however, is the Ministry of Agriculture, Fisheries and Food's guidance on Flood and Coastal Defence which is currently leading to the preparation of Shoreline Management Plans.

Statutory agencies, in particular the National Rivers Authority and English Nature, have recently been active in producing strategies which relate to their specific areas of responsibility. In the case of the NRA, a system of Catchment Management Plans is in the process of being introduced. By 1996-97, the Mersey Estuary should be covered by such a plan which can be expected to draw extensively on the Estuary Management Plan exercise.

English Nature has developed an Estuaries Initiative which has promoted the preparation of management plans for most of the major estuaries in the country. The management plans for the Dee and Ribble Estuaries are both included in this Initiative. The Draft Plan for the Dee was issued for consultation in March 1995 and that for the Ribble will be published shortly. There has been close liaison between the Study Team and the project officers responsible for the Dee and Ribble Plans to ensure that a degree of consistency is achieved between the three plans. This is especially important in the Wirral where the northern foreshore overlaps the study areas of the Dee and Mersey Plans.

International legislation is of key importance in relation to some aspects of the Plan. Particularly in the Estuary Resources section, but also in that for Economic Development, there are frequent references to European Directives which have an important influence on nature conservation and pollution control policy.

From Plan to Process

The Management Plan is an advisory document. It does not seek to replace the responsibilities of any of the agencies whose job it is to regulate the use of the Estuary. It tries to avoid adding another regulatory level for those whose livelihood depends on the Estuary. The Plan sets out to provide a framework which will help those decision-makers evaluate their actions against the wider backdrop of the Estuary as a whole.

The main purpose of the Plan is to initiate a management process which will help to ensure its measures and policies are implemented. The process will involve a number of different actors and activities. Partner organisations - those bodies that subscribe to the aims, objectives and policies of the Plan - will have a crucial role to play in establishing and developing this process. They will need to develop new policies and procedures where none exist at present and, in some cases, will need to refine and review existing approaches. Considerable effort will be needed to maintain and improve Estuary amenities and facilities and to identify new opportunities. Systematic research and monitoring will need to continue, to increase understanding of the physical and socio-economic changes that are affecting the Estuary; and it will be important to give high priority to developing mechanisms that improve consultation and coordination.

Other Relevant Material

The following material produced by the Study Team is also relevant to this chapter:

TOPIC REPORTS

1 Initial Consultation with Voluntary Organisations and the Private Sector 5 Initial Consultation with Statutory Agencies

AREA ISSUE REPORTS

1 The Upper Estuary
2 The Inner Estuary
3 The Outer Estuary

OVERVIEW REPORTS

Mersey Estuary Management Plan : First Year Report
Mersey Estuary Management Plan : Report on the Consultation Exercise.

The QE2 called at the Mersey in August 1992 on its 25th Anniversary cruise

ESTUARY RESOURCES

The overall aim of the Mersey Estuary Management Plan is to promote the sustainable use of the Estuary by ensuring that development of its economic and recreational potential is undertaken in a manner compatible with the protection, management and enhancement of the natural and man made resources of the Estuary Zone. An understanding of the nature and value of these resources and of the complex processes and interactions which continue to shape the estuarine environment is therefore fundamental to the Plan. The Estuary is a highly complex and dynamic system which is linked to the wider marine systems of Liverpool Bay and the Irish Sea and to the upper Mersey and tributary river systems. Interrelationships exist between the physical, chemical and biological components of the system, and changes in any area, whether natural or man-induced, can have impacts elsewhere. The Estuarine system has been, and continues to be, considerably influenced by human activity. The physical processes that shape the landforms and habitats within the Estuary are subject to daily and seasonal variation, as well as longer term climatic change. As a result of these changes, areas of biological importance within the Estuary can be expected to change over time. At the present time the Estuary is undergoing significant and rapid change as a result of improving water quality. In the past the highly polluted nature of the Estuary has severely restricted its ability to support wildlife. Recent improvements in water quality have, however, been reflected in an increase in the distribution and diversity of species using the Estuary, and the Mersey is now recognised as a site of international wildlife importance. Further improvements in water quality are likely to be paralleled by continued increase in biodiversity in the Estuary. As concerns about water quality have begun to ease, attention is now beginning to focus on other forms of pollution which adversely affect environmental quality in the Estuary Zone.

In addition to the dynamic nature of the Estuary itself, background studies have highlighted the need to respect the particular value of the landward areas in the Estuary Zone and the associated landscape and townscape which together provide the basis of the Region's maritime economy and identity.

These issues have informed the development of strategic objectives in the following key policy areas:

- Estuary Dynamics;
- Water Quality and Pollution Control;
- Biodiversity;
- Land Use and Development.

It is important to emphasise, however, that these areas are closely interrelated. Together they provide the context for the implementation of the Economic Development and Recreation aspects of the Mersey Estuary Management Plan.

1 : Estuary Dynamics

The Estuary environment is shaped by a complex set of physical processes which affect patterns of erosion and deposition within the Estuary Zone. These processes relate to the tide, wind and waves, and they are highly dynamic being subject to daily and seasonal variations, as well as longer-term climatic change. These processes are linked to circulation patterns in Liverpool Bay and the Irish Sea, and the Estuary system should therefore be viewed as part of a much wider and extremely complex marine system. Physical processes influence the development of landforms and habitats, and the transport and distribution of sediments and pollutants. Understanding these relationships is of fundamental importance as they influence water quality and nature conservation interest in the Estuary.

The physical processes operating within the Estuary are strongly influenced by tidal currents. The Estuary is considerably wider upstream than at its mouth, resulting in strong tidal currents through the relatively narrow entrance. Beyond the Narrows lies the expansive Inner Estuary basin, where tidal flows are slower and sand and mud are deposited to form banks which are exposed at low tide. Within this section, the low water channel shifts between the northern and southern banks, continually eroding and reworking the sediments, and preventing the growth and consolidation of intertidal banks. This process helps prevent siltation of the Inner Estuary basin, so maintaining estuarine capacity, and it plays a critical role in landform and habitat development.

The physical processes operating within the Estuary Zone will adjust to any change in the system whether natural or man-induced and it is important to recognise that the natural functioning of the system is greatly modified by human activity. Commercial navigation has been particularly influential. The development of ports and the construction of docks and canals has resulted in extensive reclamation and alteration of the natural shoreline. In addition, the nature of sedimentation within the Estuary has always caused problems for navigation, and a substantial amount of dredging is required to keep navigation channels at a constant depth.

At present, however, it is considered that the system has satisfactorily accommodated such changes, and current activities allow both the migration of the low water channel and the maintenance of a positive sediment budget which are compatible with wildlife concerns. Should sediment supply begin to fail, and/or relative sea level rise accelerate markedly, then a programme of managed set back may need to be considered to ensure the continued existence of important estuarine habitats. Equally, if there was a marked increase in sediment supply, not only would costly dredging requirements increase, but current intertidal landforms, particularly high salt marsh, could become permanent dry land with implications both for land claim proposals and, in such a polluted environment, the risk of increased pollutant uptake into food chains.

Controlling human activity to ensure that the Estuary continues to function as naturally as possible and in a self sustaining way is considered to be a primary objective of the Management Plan.

OBJECTIVE   *To allow the Estuary to function as naturally as possible and in a self sustaining way by controlling human interference in intertidal and marine areas having regard to the natural conditions and processes of the Estuary and Liverpool Bay.*

POLICIES

MEMP and Specialist Studies

Hydraulic studies of the Mersey have shown how past engineering works in the Estuary Zone have altered circulation patterns in Liverpool Bay and restricted the movement of the low water channel. Activities such as land reclamation, changes in dredging patterns, the disposal of dredging spoil and aggregate extraction all have the potential to alter the physical regime of the Estuary with possible adverse effects on nature conservation, fisheries, navigation and recreation interests.

Under present statutory arrangements no activity with the potential to affect the dynamics of the Estuary can be undertaken without the prior consent of the Acting Conservator of the River Mersey. This involves the insistence on specialist studies where existing information and expertise are considered to be inadequate. The Acting Conservator also has powers to instigate studies and require remedial action to be undertaken where it is thought that an existing activity has resulted in adverse changes in Estuary dynamics. The Acting Conservator has particular responsibilities in relation to the protection of navigation interests but also exercises a watching brief on general Estuary concerns.

The role of the Acting Conservator is complemented by the FEPA (Food and Environmental Protection Act 1985) licensing system operated by the Ministry of Agriculture, Fisheries and Food (MAFF). This seeks to ensure the protection of the marine environment including its living resources, and requires licence applications to be made for marine construction projects and the disposal of dredged spoil. MAFF may also require specialist studies to be undertaken to assess the impact of licence applications in certain circumstances. In determining the need for, and content of, specialist studies related to Estuary dynamics and in assessing their results it is proposed that specific consideration should be given to the full spectrum of concerns set out in the Management Plan.

The Mersey has been relatively fortunate in the extent of information available on Estuary dynamics and this has enabled informed judgments to be made about the need for additional studies. This situation has arisen mainly as a result of development pressures, such as the proposed Mersey Barrage, and the operational requirements of commercial navigation interests in the Estuary.

Field measurements of currents and wave energy, and models of estuarine circulation have improved understanding of the physical processes operating within the system. Tide gauge data is collected from four locations within the Estuary. In addition bed levels throughout the tidal Mersey were surveyed systematically by Mersey Docks and Harbour Board/Company from 1861 to 1977. Since 1977 very few significant surveys, and none covering the entire tidal limits, have been carried out.

Without comprehensive bed level data it is not possible to maintain a grasp of the prevailing relationship between inter-tidal banks and channels, nor can calculations be made of tidal capacity which is the most effective means of monitoring the impact of natural and man-induced events on the health of the Estuarine system as a whole. Background studies have also suggested the need for an up-to-date database of the patterns and rates of process operation within the Estuary, coupled with a similarly up-to-date inventory of the resource base. The need to establish regular and comprehensive Estuary surveys must be addressed in the implementation of the Management Plan and it is suggested that an early review of the existing information relating to the physical dynamics of the Estuary is undertaken. This should determine the adequacy and availability of existing data sets to predict future changes (including the potential effects of climatic change) identify any gaps in information coverage, and establish long term monitoring requirements.

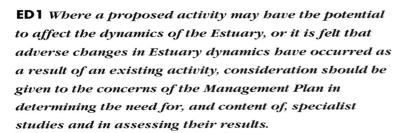

ED1 *Where a proposed activity may have the potential to affect the dynamics of the Estuary, or it is felt that adverse changes in Estuary dynamics have occurred as a result of an existing activity, consideration should be given to the concerns of the Management Plan in determining the need for, and content of, specialist studies and in assessing their results.*

Management Measure

ED1.1 *Reviewing existing research and monitoring programmes related to physical processes in the Estuary to determine the adequacy and availability of existing data sets, identify gaps in information coverage, and establish long term monitoring requirements.*

Estuary Dynamics and New Development

Partner organisations need to be aware that new developments within and beyond the Estuary Zone have the potential to adversely affect the semi-natural functioning of the Estuary. These effects include various issues related to flood risk. For example, development on flood plains is at risk from flooding, and can increase the risk of flooding elsewhere, by reducing the storage capacity of the flood plain, so impeding flood flows. Development elsewhere in the catchment may increase surface water run-off, so adding to the risk of flooding further downstream, and may also increase the risk of pollution and damage to estuarine habitats. Similarly, development which threatens the stability and continuity of fluvial and tidal flood defences can place large areas at risk from inundation. Partner organisations should take account of all aspects of Estuary dynamics including those related to flood risk in considering development

proposals. Areas currently considered to fall within a 1:200 year flood envelope are shown on Map 1. These areas include those protected by existing sea defences. Partner organisations should pay particular attention to new development proposals within these areas.

ED2 *In undertaking their activities partner organisations should have regard to the importance of restricting development which would:*

(i)   adversely affect the ability of the Estuary to function as naturally as possible in a self sustaining way; or

(ii)   increase the risk of flooding or coastal erosion through its impact on natural coastal processes; or

(iii)   prejudice the capacity of the coast to form a natural sea defence or adjust to changes in conditions, without risk to life or property; or

(iv)   be located in areas at risk from coastal flooding, erosion, or landslip, where this would increase the need for additional sea walls or other engineering works for coast protection and flood defence purposes except where necessary to protect existing investment.

Coastal Defence : Good Practice

The Ministry of Agriculture, Fisheries and Food aims to ensure that, where there is a need for protection against flooding or coastal erosion, this should be provided in the most environmentally sensitive manner and that the option of using natural coastal systems should be favoured wherever possible. MAFF is promoting such an approach in their *Strategy for Flood and Coastal Defence in England and Wales* and documents such as *Coastal Defence and the Environment : A Guide to Good Practice.* It is suggested that the principles set out in these documents should guide coastal defence activities within the Estuary Zone.

ED3 *In considering coastal protection and flood defence requirements, partner organisations should seek to:*

(i)   ensure that the works are necessary to protect life, existing built development or fixed capital assets which cannot be relocated inland;

(ii)   encourage habitat creation through the use of 'soft engineering techniques', including managed set back, wherever these provide a realistic alternative to the maintenance of 'hard defences';

(iii)   make appropriate allowance for sea level rise;

(iv)   ensure that there are no unacceptable impacts on water quality, wildlife, coastal features, landscape quality and the development and recreation concerns of the Management Plan;

(v)   ensure that satisfactory measures are undertaken to minimise the environmental impact of any construction activity.

Links to Shoreline Management Plans and the Liverpool Bay Coastal Cell

One of the important developments highlighted in the MAFF *Strategy for Flood and Coastal Defence in England and Wales* is a requirement for coastal authorities to prepare Shoreline Management Plans. These plans are designed to encourage effective management of the coastline and, according to MAFF guidelines, should cover issues related to coastal defence, opportunities for development and conservation and the identification of associated research and monitoring requirements. Within the Estuary Zone, Wirral and Sefton Metropolitan Borough Councils are involved in the preparation of Shoreline Management Plans and it is important that these documents reflect the concerns of the Estuary Management Plan and can benefit from any future research and monitoring arrangements that may emerge.

Effective coordination of individual Shoreline Management Plans is also important as coastal processes operate over large geographical areas and shoreline alterations, such as coastal defence works, can have impacts elsewhere within the system. A series of "Coastal Cell" areas have been defined nationally based on the circulation of coarse sediments. Within these areas, representatives from the various agencies with responsibilities for coast protection are coming together to ensure that a co-ordinated and effective approach to Shoreline Management Plans is achieved. The Liverpool Bay Coastal Cell extends from Llandudno to the Ribble Estuary and includes all of the Wirral and Sefton shorelines. In addition to the two Merseyside authorities, agencies represented on the Coastal Cell Group include Delyn, Ruddlan, Colwyn and Aberconwy Borough Councils, the NRA (North West Region) and the NRA (Welsh Region). The Group is to prepare a Shoreline Management Strategy which will guide the preparation of Shoreline Management Plans for the three sediment sub-cells within its area. These cover the Sefton and West Lancashire coast, the Wirral coast and the North Wales coast between the River Dee and Great Orme. Again it is important that the Shoreline Management Strategy for the Coastal Cell reflects the concerns of the Mersey Estuary Management Plan as well as the adjoining Estuary Management Plans being prepared for the Dee and the Ribble.

ED4 *Partner organisations should ensure compatibility between the Mersey Estuary Management Plan and shoreline management activities within the Liverpool Bay Coastal Cell.*

Coastal Defence within the Estuary Zone

Arrangements for the effective coordination of shoreline management activities are developing for the area falling within the Liverpool Bay Coastal Cell. The definition of the Cell, however, excludes significant areas of the Inner and Upper Mersey Estuary which are linked to the marine systems within Liverpool Bay and where coast and flood defence issues are also significant. The NRA (North West Region) chair a Merseyside and Cheshire Coast Protection and Sea Defence Liaison Committee, but in general coordination of shoreline management issues is much less formal outside the Coastal Cell. It is important that partner organisations also ensure coordination of coast defence activities within the Inner and Upper Estuary and that appropriate links are maintained with the Liverpool Bay Coastal Cell Group.

ED5 *Partner organisations should seek to coordinate coast defence activities within the Estuary Zone and maintain links with the Liverpool Bay Coastal Cell Group. In undertaking these activities regard should be paid to the concerns of the Mersey Estuary Management Plan.*

2 : Water Quality and Pollution Control

The Mersey Estuary drains a highly urbanised and industrialised catchment and has been used as a means for the disposal of industrial and domestic waste since the time of the Industrial Revolution. This has resulted in gross pollution and some sections have been among the most polluted waters in Britain. On the basis of the National Water Council Classification Scheme, the Inner Estuary is currently categorised as Class 3 (poor) and the Upper Estuary Class 4 (bad). The present classification scheme is based on dissolved oxygen levels, aesthetics and biological conditions. These gradings are regarded as being rather subjective and a completely new scheme is currently being developed by the NRA for future use.

The polluting materials entering the Estuary originate from two main sources, industrial effluents and sewage discharges. These discharges have a high biochemical oxygen demand and include a variety of persistent materials. Organic matter is broken down by bacteria, so depleting the water column of dissolved oxygen thereby affecting marine life. Heavy metals discharged into the Estuary are adsorbed onto estuarine sediments or transported into Liverpool Bay either as suspended particles or as dredged spoil. The long-term discharge of industrial effluents to the Estuary and the persistence of these materials has meant that Estuarine sediments are contaminated with heavy metals such as mercury, lead, copper, chromium and zinc.

Water quality in the Estuary has, however, improved considerably in recent years as a result of initiatives such as the Mersey Basin Clean Up Campaign and more stringent EU requirements. Major improvements to sewage treatment works have been undertaken by North West Water PLC in order to reduce pollution loads and, although not fully completed, there have been a number of significant achievements. For example, the

The new treatment works at Sandon Dock, Liverpool, which replaced 28 crude sewage discharges along the Estuary's northen shore

pollution load in rivers flowing into the Estuary has been reduced by 80% and mercury discharges have been reduced by more than 90% over the last 15 years. Further improvements can be anticipated when current improvements to treatment works are completed and secondary treatment facilities are introduced in line with the EU Urban Waste Water Directive over the next few years. The Mersey Basin Campaign hopes to achieve Class 2 (fair) status for the entire Estuary by the year 2010. This means that the Mersey should be able to support a reasonably diverse biological life including some sustainable fish life, and be aesthetically acceptable (i.e. the discharge of crude sewage should have ceased). In order to achieve this target River Ecosystem Water Quality Objectives are being set for all upstream rivers.

A recent review of environmental quality in the Estuary undertaken by the NRA suggests that the Mersey is well on the way to achieving the Mersey Basin Campaign target and, although water quality is likely to remain unsatisfactory for some time to come, it is considered that the tide of water pollution has now turned and real improvement in water quality is well established. In these circumstances attention is now increasingly focusing on other forms of pollution which adversely affect environmental quality in the Estuary Zone. For example, concerns about air quality have recently been highlighted as a result of two proposed incinerator projects designed to reduce polluting discharges into the Estuary. Similarly the industrial heritage of the area has created a legacy of highly contaminated sites adjoining the Estuary and some current industrial activities, such as coal handling in the Dock Estate, create pollution problems which seriously detract from local environmental quality.

Continuing improvement in pollution control is considered to be a major objective of the Management Plan and partner organisations should work together to achieve an integrated approach to pollution issues and environmental good practice within the Estuary Zone.

OBJECTIVE *To support continuing improvements in water, air, land, noise and light quality and the adoption of environmental good practice within the Estuary Zone.*

POLICIES

Pollution Control : Best Practice

Under current legislation, environmental quality objectives and standards are set to ensure that safe levels of particular substances are maintained and that no further increase in the concentration of toxic substances in the environment occurs. The Plan aims to ensure that these standards are met, and to encourage organisations to adopt the highest possible standards of pollution control in excess of current standards wherever feasible. In order to improve environmental performance it is recommended that partner organisations should consider the adoption of British Standard 7750 related to Environmental Management Systems and appropriate Environmental Codes of Practice and actively promote the benefits of such approaches within the Estuary Zone.

In addition to considering environmental performance at an organisation level, partner organisations should adopt an integrated approach to pollution control matters more generally within the Estuary Zone. This should include the establishment of agreed pollution reduction targets where they do not already exist. In this context specific consideration should be given to the adoption of targets related to EU Sea Bathing Standards in some areas and for water quality improvements related to the needs of the fishing and shellfishing industries. MAFF have, for example, expressed particular concern about water quality standards in the vicinity of the commercial cockle and mussel beds at Crosby and on the North Wirral Coast (see Map 1) and are keen to promote the maintenance and improvement of the water quality in these areas. Joint working to achieve targets may be necessary and particular assistance should be given to North West Water PLC to enable it to carry out its pollution reduction programme. This should include consideration of the land requirements of waste water treatment and disposal. Partner organisations should also establish an ongoing programme of monitoring to check progress towards agreed pollution reduction targets.

PC1 *Partner organisations should promote the adoption of best practice in order to continue to reduce water, air, land, noise and light pollution within the Estuary Zone and should work together to meet agreed pollution reduction targets.*

Management Measures

PC1.1 *Promoting discussion, information exchange and the dissemination and adoption of best practice in relation to pollution control within the Estuary Zone.*

PC1.2 *Establishing agreed pollution reduction targets within the Estuary Zone. The case for action to achieve EU Sea Bathing standards and water quality improvements related to the needs of the fishing and shellfishing industries should be specifically addressed in this context.*

PC1.3 *Assisting North West Water PLC in their efforts to meet pollution reduction targets. This should include consideration of the land requirements for waste water*

treatment and disposal.

PC1.4 *Reviewing existing monitoring programmes to determine what additional work is required and supporting an appropriate programme of ongoing monitoring derived from clear objectives to check progress towards defined targets.*

Pollution Control and New Development

New development can have significant effects on the quality of surface, underground and coastal waters and continuing vigilance is necessary to maintain the trend of water quality improvement in the Estuary. Particular care is needed to protect groundwater sources, such as those in Wirral, as pollution of such sources is difficult and expensive to remedy. The supply of water to new developments is also significant. Development should not normally be permitted where water supplies are inadequate or where new supply requirements may adversely affect the water environment. It is also important to ensure that foul sewers and treatment works of adequate capacity and design are available or will be provided in time to serve the development. Similarly new development may cause air, noise and light pollution which may impinge on local environmental quality and, in some instances, have more extensive environmental consequences.

Partner organisations should give careful consideration to any development likely to give rise to pollution problems or adversely affect the achievement of pollution reduction targets. Proposals related to mineral extraction, waste disposal and industrial or chemical processes or located on contaminated sites may carry particular risks. Where satisfactory pollution control measures cannot be adopted, developments should be resisted. Particular attention should be given to the risk of damage and/or disturbance in environmentally sensitive areas and areas of quiet recreation within the Estuary Zone.

PC2 *In considering development proposals partner organisations should have regard to the importance of ensuring that:*

i) *adequate water supplies exist to service the development;*

ii) *there are adequate facilities for the treatment and disposal of additional surface and waste water discharges resulting from the development;*

iii) *appropriate pollution prevention measures are adopted in areas of groundwater vulnerability;*

iv) *appropriate measures are undertaken to prevent water, air, noise, land, light pollution both during and after construction and that special regard is given to the risk of damage or disturbance to environmentally sensitive areas (see Policies BD1 and BD2) and areas of quiet recreation.*

Contaminated Land

The industrial heritage of the Mersey Estuary has created a legacy of land contamination in significant areas of the Estuary Zone, most notably in the Inner and Upper Estuary. This contamination has implications for water quality in the Estuary, for nature conservation interests and for development potential. Partner organisations should promote a coordinated programme of care and remedial action for contaminated land within the Estuary Zone which should include consideration of the most appropriate long term use of sites. In some cases it may be appropriate to leave sites substantially untouched as disturbance may cause severe and unpredictable pollution impacts. In these instances

built development may be inappropriate. As an initial step, a survey and review of contaminated land should be undertaken to assess the options for remedial action and long term use.

PC3 *Partner organisations should promote a coordinated programme of care and remedial action for contaminated land in the Estuary Zone.*

Management Measures

PC3.1 *Undertaking a survey and review of contaminated land within the Estuary Zone and assessing the options for the application of remedial measures and long term use.*

PC3.2 *Establishing a coordinated programme of care and remedial treatment for contaminated land within the Estuary Zone.*

MEMP and Water-related Initiatives

The Mersey Estuary Management Plan is only one of a number of important water-related initiatives currently being developed within the region. The NRA, for example, is in the process of producing a new series of Catchment Management Plans which set out strategies for planned improvement in the water environment associated with individual river systems. The Estuary Management Plan will provide a valuable basis for the preparation of the Mersey Estuary Catchment Management Plan which is due to be published in 1996/97. When complete both documents should complement each other and bring together the management of all water-based interests in and surrounding the Estuary. It is also important to recognise that the Estuary is closely linked to both the upper reaches of the Mersey and to tributary systems including the River Alt and the River Weaver. In addition to the Mersey Estuary Catchment Management Plan, similar documents are being, or will be, prepared for the Upper Mersey, the Middle Mersey, the Alt and Crossens and the Weaver. The interactions between these areas and plans and the Mersey Estuary should be considered carefully.

In parallel with these exercises a number of River Valley Initiatives are being undertaken as part of the Mersey Basin Campaign. These initiatives bring Campaign resources to bear on specific locations and offer a focus for local involvement in river-related environmental projects. The Alt 2000 RVI which was established in 1992 is particularly significant in the context of the Mersey Estuary Management Plan and it is important that effective links between the two exercises are maintained. The Alt 2000 RVI has been very successful in attracting the involvement of all sectors of the community in improvement projects and it is felt that the Mersey Estuary Management Plan could beneficially draw upon the ideas and expertise developed through this work. A number of new River Valley Initiatives are being considered and links with the Estuary Management Plan should be should be developed where appropriate.

PC4 *Partner organisations should seek to ensure appropriate coordination between the Mersey Estuary Management Plan and the forthcoming Mersey Estuary Catchment Management Plan and in other Catchment Management Plans and River Valley Initiatives related to the Mersey Estuary.*

3 : Biodiversity

The continual input, trapping and recycling of sediments and nutrients make estuaries amongst the most productive ecosystems in the world. Their uniquely varying set of physical, chemical and biological conditions creates a complex mosaic of subtidal, intertidal and surrounding terrestrial habitats which support abundant and varied plant and animal communities.

Despite its legacy of pollution, the Mersey Estuary provides a variety of valuable estuarine habitats including dune systems, intertidal flats, rocky shores, salt marsh and permanent water including dock features. These habitats support a variety of associated species, and the Estuary Zone is recognised as being of particular international importance in relation to over-wintering and passaging birds.

Water pollution affecting the Estuary has, however, suppressed species diversity and experience suggests that, as water quality is improving, the number and spatial distribution of species supported by the Estuary are increasing. Continued improvements in water quality are expected to be paralleled by further increases in species diversity and it is likely that that this will be accompanied by the relative decline of some species as the nature of the Estuary changes and it moves towards a species profile that is more akin to that of an undeveloped estuary. While conservation of established habitats and species should be given high priority this should be undertaken sensitively and should not be at the expense of the process of transition towards a more natural system.

OBJECTIVE *To conserve, and where relevant restore, the natural biodiversity of the Estuary Zone.*

POLICIES

Site and Habitat Protection

Many parts of the Estuary Zone already attract special protection as being of local, national or international nature conservation importance. Sites currently the subject of statutory designations are shown on Map 1. These sites include nationally important Sites of Special Scientific Interest (SSSIs) located within the Inner Estuary, on the North Wirral Foreshore, at Formby Sand Dunes and Foreshore and at Woolston Eyes. In terms of international designations, Altcar Sand Dunes and Foreshore have attracted RAMSAR and Special Protection Area (SPA) status and both designations are proposed for the Mersey Estuary SSSI. The adoption of the EU Directive on the Conservation of Natural Habitats and of Wild Fauna and Flora ("The Habitats Directive") requires the UK Government to identify Special Areas of Conservation (SACs) and the appropriateness of such designations within the Estuary Zone should be considered.

It should be noted that site designations may change over time. For example, boundaries of existing designations may be altered to reflect changes in the ecology of the area, some sites may become less important over time and be de-designated, other sites of importance may emerge. One example of the latter relates to the current identification of Regionally Important Geological/ Geomorphological Sites which, although not strictly an aspect of biodiversity, are of nature conservation note and are in the process of acquiring formal status by designation in development plans. The present position as far as these sites are concerned is shown in Map 1.

The Management Plan fully supports the protection of sites attracting formal nature conservation designations within the Estuary Zone. Partner organisations should give the highest degree of protection available under the appropriate legislation to these sites and avoid any action which may adversely affect, either directly or indirectly, their conservation status. The need to consider indirect effects is particularly important within the context of the Estuary given the nature of Estuary dynamics and the sensitivity of many of the formally designated sites to disturbance from outside the area. In recognition of the need to pay special regard to nature conservation interests in areas surrounding key sites, English Nature is in the process of identifying consultation zones around SSSIs.

In order to assess the effectiveness of the Management Plan in providing appropriate protection, a baseline survey of habitats and sites of nature conservation value, including those of geological and geomorphological value, needs to be undertaken. Much is already known about the distribution of habitats within the Estuary Zone. The baseline survey should draw upon established sources, such as the Mersey Barrage studies, as far as possible, and identify and fill the most significant gaps.

In undertaking the baseline survey, consideration should be given to the need for additional formal designations, particularly in the Upper Estuary, having regard not only to individual merit, but also to the contribution of sites to the overall ecology of the Estuary.

The baseline survey should be used to inform the development of a programme of on-going monitoring and in this context partner organisations should consider establishing safeguarding measures for key habitats. These would identify levels below which certain habitats should not fall and, in recognition of the dynamic nature of the Estuary, should generally be phrased in a non-site-specific way. Consideration should also be given to the definition of enhancement targets if the extension of certain habitat types is thought to be desirable.

Safeguarding measures and enhancement targets would provide a basis on which to measure the effectiveness of the plan, assist in the consideration of individual development proposals and act as triggers for conservation action. Care must, however, be taken in utilising safeguarding measures and targets so as to accommodate "uncontrollable" changes in the Estuary system such as those caused by sea-level rise and support the general improvement in the ecology of the Estuary.

Efforts to protect key sites and habitats may be usefully complemented in some instances by the introduction of supporting management practices. Detailed site management plans enable owners, occupiers and other users to manage land areas in a manner compatible with nature conservation interests. Partnership agreements with owners, occupiers and others have already been established for some sites within the Estuary Zone, most notably along the Sefton Coast. Consideration should be given the desirability of extending such arrangements to other areas and a prioritised programme of site management plan preparation and review should be established. This programme should clearly identify lead organisations and sources of finance in each case. The importance of the water should be recognised, and management plans should be extended to cover the water areas, and water-based issues where they pose a threat to the nature conservation value of a site. In assessing the need for site management plans specific consideration should be given to the case for introducing controls over the spread of Spartina grass in the Mersey Estuary.

BD1

Partner organisations should give the highest degree of protection available under the appropriate legislation to sites and habitats within the Estuary Zone which are recognised as being of international, national and/or local importance.

Management Measures
BD1.1 *Undertaking a baseline survey of sites and habitats in the Estuary Zone and establishing an appropriate programme of on-going monitoring.*

BD1.2 *Establishing safeguarding measures and, where appropriate, enhancement targets in relation to key habitats and natural features. These safeguarding measures should be respected by partner organisations in undertaking their activities.*

BD1.3 *Preparing Management Plans, where appropriate, for sites and habitats covered by BD1 in order to maintain their favourable conservation status.*

Species Protection
The area designations within the Estuary Zone reflect their importance in relation to particular species which are rare or endangered or, as in the case of a number of bird species, are present in nationally and internationally important numbers. Habitat protection will in most cases secure the protection of such species but it is considered that separate identification of key species may provide a desirable level of added protection in some instances. Nature conservation agencies, MAFF, landowners, and other interest groups should identify which species merit specific protection in the context of the Estuary Zone and partner organisations should give them the highest degree of protection available under the appropriate legislation.

In order to inform this process and establish on-going monitoring requirements, a baseline survey should be undertaken of the presence, distribution and numbers of species present within the Estuary Zone which are recognised as being of international, national or local importance.

Species-related safeguarding measures and enhancement targets should be developed where appropriate. These may not only relate to the numbers and distribution of species in the Estuary but also relate to the relative health of individual species. The need for this type of additional measure of nature conservation performance has recently been highlighted by research undertaken on behalf of the NRA which revealed raised levels of certain chemicals in the muscle tissue of fish caught in the Estuary. In devising safeguarding measures and enhancement targets for individual species it is important that the changing character of the Estuary is fully recognised and that progress towards a more natural system is accommodated.

Consideration should also be given to the need to complement site management plans with the preparation of action plans for individual species. This may be particularly appropriate in relation to certain bird species which depend for their survival on sites not covered by formal nature conservation designations. Disturbance from human activities such as shooting and motorbiking can be extremely disadvantageous to many species and control of such activities in particular areas and/or at particular times of the year could greatly assist nature conservation objectives.

BD2 *Partner organisations should give the highest degree of protection available under the appropriate legislation to those species present within the Estuary Zone which are recognised as being of international, national and/or local importance.*

Management Measures

BD2.1 *Undertaking a baseline survey of presence, distribution and numbers of species in the Estuary Zone and establishing an appropriate programme of on-going monitoring.*

BD2.2 *Establishing safeguarding measures and, where appropriate, enhancement targets in relation to key species. These safeguarding measures should be respected by partner organisations in undertaking their activities.*

BD2.3 *Preparing Action Plans, where appropriate, for species covered by BD2 in order to maintain and, wherever possible, improve their conservation status.*

Creative Conservation

While the protection of key habitats and species is a primary concern of the Mersey Estuary Management Plan this may be usefully complemented by the promotion of creative conservation practices in some instances. Many areas of the Estuary Zone are not of great current nature conservation value in their own right but do contribute in some way to the overall nature conservation value of the area. In other instances the nature conservation value of sites could be substantially increased by sympathetic development, redevelopment or the adoption of appropriate estate management practices. In addition, there may be circumstances where the loss of a significant species or habitat is unavoidable and mitigation measures must be considered. It should be noted, however, that if mitigation measures are proposed great care should be taken to ensure that any replacement is of an equivalent type, extent and quality

View over Ince Banks, part of the Site of Special Scientific Interest which covers most of the Inner Estuary

to that which is being lost and should be located as close to the original site as possible.

BD3

Partner organisations should promote the adoption of creative nature conservation practices. Where possible in the Estuary Zone new development, redevelopment, estate management and farming practices should:

i) *retain woodland, trees, hedgerows, watercourses, ponds and other established habitats, and geological and geomorphological features of conservation interest. Particular attention should be given to minimising water level changes and soil compaction;*

ii) *make compensatory provision for replacement habitats / features of equivalent type and quality and of equivalent or greater extent where the loss of existing habitats/features is unavoidable;*

iii) *incorporate, wherever appropriate, the creation of new wildlife habitats and fresh geological exposures and relate these to neighbouring sites in order to contribute to the development of wildlife corridors;*

iv) *incorporate the greatest possible proportion of appropriate native planting in any landscaping or planting scheme;*

v) *incorporate, where appropriate, the maximum possible level of permeable ground surface and take such steps as are necessary to regulate surface water flows from impermeable surfaces;*

vi) *open up culverted watercourses and restore the continuity of riverine corridors.*

4 : Land Use and Development

Government guidance set out in PPG20 on Coastal Planning highlights the growing appreciation of the coast as an important national resource. The guidance emphasises the need to respect the particular value of the coast as a location for a range of economic and social activities and to protect the special character of natural and historic landscapes in coastal areas.

These concerns are very relevant to the Mersey Estuary Management Plan. The banks of the Mersey have provided the context for the development of the Region's maritime economy and for a range of major industrial and other uses which have benefited from a waterside location. Development of the river- banks has been extensive and the built environment contains magnificent examples of fine engineering and architecture, with a rich maritime history. However, a few areas of open coast and natural shoreline still remain and these areas make an important contribution to the overall environmental quality and character of the Estuary. Careful stewardship of land resources in the Estuary Zone and protection and enhancement of the Estuary's landscape and townscape is an important aspect of the Mersey Estuary Management Plan.

OBJECTIVE *To promote careful stewardship of land resources, landscape and townscape within the Estuary Zone.*

POLICIES

Development within the Estuary Zone

The Estuary's riverbanks and inter-tidal areas have been subject to historic development for many uses including docks, airports, power stations, industry, storage and distribution depots, bridges, canals, waste disposal sites and housing. As a result, the Mersey is among the most highly developed estuaries in Britain with few remaining areas of open coast or natural shoreline. These remaining areas are of particular significance for a variety of reasons. Such areas provide relief from the highly urbanised and industrial landscape of the Estuary and are frequently important as venues for informal recreation. Some sites also provide valuable wildlife habitats or contain features of significant geological or geomorphological interest. In addition, the construction of sea walls and other coast defence structures has increased the importance of the remaining stretches of natural shoreline particularly in light of possible rises in sea level. Further encroachment of the areas of open coast indicated on Map 1 should be avoided and sites of geological, geomorphological or biological interest, or which form an existing or potential sea defence, should be given the highest degree of protection from development. Conversely, new development should be concentrated in the existing developed areas of the Estuary Zone.

Central government guidance indicates that uses and activities requiring a coastal location include, for example: tourism; recreation; ports; marinas; industries importing bulky raw materials that depend on access to the sea; mineral extraction; energy generation; and waste water and sewage treatment and disposal. Within the developed areas of the Estuary Zone partner organisations should have regard to the particular needs of such uses and in the immediate waterfront area should give priority to developments requiring a coastal location.

A coordinated approach to land use and development within the Estuary Zone is important and when reviewing their development plans and informal plans, local authorities should give consideration to the identification of a coastal/estuary/river valley corridor or zone with a view to achieving consistency in terms of policies, proposals and presentation.

Given the objectives of the Mersey Estuary Management Plan and its concern to protect the remaining areas of open coast, a review of the continuing relevance of industrial land allocations and the need for the Secretary of State for the Environment's special direction related to the Mersey Marshes Local Plan area should be undertaken.

LU1

Partner organisations should encourage new development, especially that requiring an estuary location to locate within the existing developed areas of the Estuary Zone. Development should be discouraged from locating within the remaining areas of open coast. Sites of geological, geomorphological or biological interest, or which form an existing or potential natural sea defence, should be protected from development.

Management Measures

LU1.1 *When reviewing their development plans and informal plans, local authorities should give consideration to a coordinated approach to identifying a coastal/estuary/river valley corridor or zone with a view to achieving consistency in terms of policies, proposals and presentation between different sections of the Estuary Zone.*

LU1.2 *Undertaking a review of the continuing relevance of the industrial land allocations set out in the Mersey Marshes Local Plan.*

Built Environment

Special attention should be paid to the quality of the built environment within the Estuary Zone and partner organisations should take steps to protect and enhance this wherever possible. This can be achieved partly by the careful control of new developments. Dramatic waterfront skylines are a feature of the Estuary and, where new development is acceptable in locational terms, every effort should be made to ensure that the scale and form of developments is sympathetic to its surroundings and that it does not detract from key views of the Estuary and its setting, and from the Estuary itself.

To assist partners in these efforts an assessment should be undertaken to describe, classify and evaluate the various components of the townscape within the Estuary Zone. This should inform the development of townscape guidelines and an associated programme of conservation and enhancement work.

LU2 *Partner organisations should seek to protect and enhance the quality of the built environment adjoining the Estuary. The form and scale of new development which would be visible from viewpoints along the waterfront and from the River should respect the special character of the Estuary.*

Management Measures

LU2.1 *Undertaking an assessment to describe, classify and evaluate the various components of the townscape in the Estuary Zone.*

LU2.2 *Developing appropriate townscape guidelines for new development within the Estuary Zone and a programme of associated conservation and enhancement work.*

Retaining Waterfront Heritage

The Estuary benefits from a strong maritime heritage which gives it a distinctive identity. This heritage should be respected and wherever possible partner organisations should seek to retain and enhance existing docks, waterspaces and historic buildings and features and their settings. Existing Conservation Areas and the sites of Ancient Monuments are shown in Map 1. Attention should be directed not only to large scale features but should also encompass small scale artifacts wherever practical. There are already many examples of effective reuse of historic buildings and features within the Estuary Zone and this experience should be drawn upon to promote reuse of heritage assets wherever possible.

LU3 *Wherever possible partner organisations should seek to retain the Estuary's waterfront heritage, including the existing docks and waterspaces and historic buildings and features and their settings. Opportunities for enhancement and appropriate reuse should be promoted especially where these may help regeneration.*

Promoting Landscape Quality

In undertaking their activities partner organisations should also pay particular attention to the landscape character of the open coast and adjacent areas. The character of areas of high landscape value, such as the dune coasts of Sefton and Wirral and within the Inner Estuary the area

Area of open coast at Hale Head

around Hale Head, should be protected and opportunities for landscape enhancement should be promoted generally within the Estuary Zone.

To assist partners in these efforts an assessment should be undertaken to describe, classify and evaluate the various components of the landscape within the Estuary Zone. Such a study should draw upon the work already undertaken as part of the preparation of the Mersey Forest Plan. This should inform the development of landscape guidelines for the Estuary Zone and an associated programme of conservation and enhancement work.

LU4 *The character of the open coast and adjacent areas should be preserved and, where appropriate, opportunities for landscape enhancement and informal recreation should be promoted.*

Management Measures

LU4.1 *Undertaking an assessment to describe, classify and evaluate the various components of the landscape in the Estuary Zone.*

LU4.2 *Developing appropriate landscape guidelines for new development within the Estuary Zone and a programme of associated conservation and enhancement work.*

Figure 4: Estuary Resources: Summary of Policy Areas, Policies and Management Measures.

Strategic Policy Area	Strategic Policy		Management Measure	
1 ED: Estuary Dynamics	**ED1:**	*MEMP and Specialist Studies*	**ED1.1:**	Review of Monitoring Programmes
	ED2:	*Estuary Dynamics and Development*		
	ED3:	*Coastal Defence Good Practice*		
	ED4:	*Links to Liverpool Bay Coastal Cell*		
	ED5:	*Coastal Defence within the Estuary Zone*		
2 PC: Water Quality and Pollution Control	**PC1:**	*Pollution Control Best Practice*	**PC1.1:** **PC1.2:** **PC1.3:** **PC1.4:**	Promoting and Disseminating Best Practice Agreeing Pollution Reduction Targets Meeting Pollution Reduction Targets Agreeing Programme of Monitoring
	PC2:	*Pollution Control and Development*		
	PC3:	*Contaminated Land*	**PC3.1:** **PC3.2**	Survey and Assessment of Options Programme of Care and Remedial Treatment
	PC4:	*MEMP and Water-related Industries*		
3 BD: Biodiversity	**BD1:**	*Site and Habitat Protection*	**BD1.1:** **BD1.2:** **BD1.3:**	Baseline Survey Safeguarding Measures and Enhancement Targets Plans for Sites, Habitats and Features
	BD2:	*Species Protection*	**BD2.1:** **BD2.2:** **BD2.3:**	Baseline Survey Safeguarding Measures and Enhancement Targets Plans for Species
	BD3:	*Creative Conservation*		
4 LU: Land Use and Development	**LU1:**	*Development within the Estuary Zone*	**LU1.1:** **LU1.2:**	Coordinating Plans within Corridors Reviewing Land Allocations for Industry
	LU2:	*Built Environment*	**LU2.1:** **LU2.2:**	Assessment of Townscape Developing Townscape Guidelines
	LU3:	*Retaining Waterfront Heritage*		
	LU4:	*Promoting Landscape Quality*	**LU4.1:** **LU4.2:**	Assessment of Landscape Developing Landscape Guidelines

Other Relevant Material

The following material produced by the Study Team is also relevant to this chapter:

TOPIC REPORTS

1 Initial Consultation with Voluntary Organisations and the Private Sector.

2 Navigation, Tidal Regime and Land Use.

3 Water Quality and Nature Conservation.

5 Initial Consultation with Statutory Agencies.

6 Landownership and Tenure.

7 The EC Waste Water Treatment Directive.

10 Emergency Planning.

11 Fishing and the Mersey Estuary.

12 Coast and Flood Defence.

OVERVIEW REPORTS

Mersey Estuary Management Plan : First Year Report.

Mersey Estuary Management Plan : Report on the Consultation Exercise.

MAP 1 ESTUARY RESOURCES

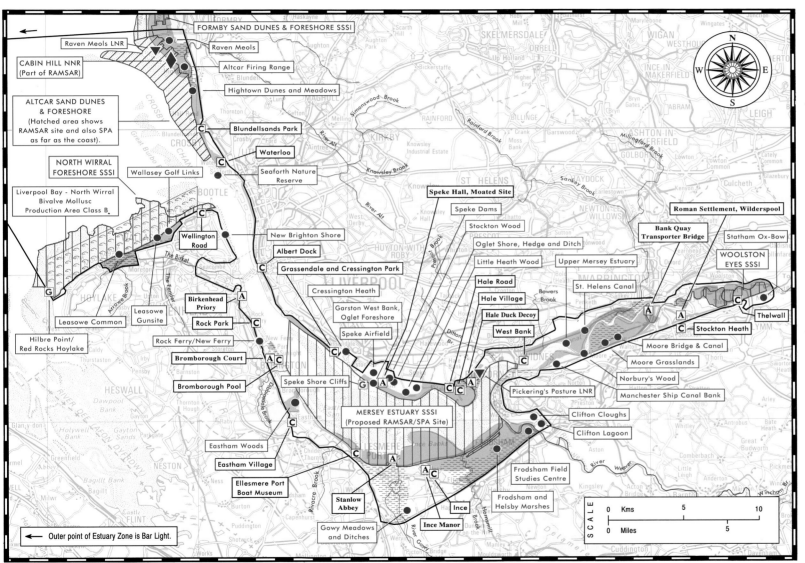

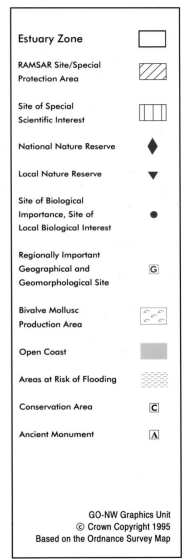

Estuary Zone

RAMSAR Site/Special Protection Area

Site of Special Scientific Interest

National Nature Reserve ◆

Local Nature Reserve ▼

Site of Biological Importance, Site of Local Biological Interest ●

Regionally Important Geographical and Geomorphological Site G

Bivalve Mollusc Production Area

Open Coast

Areas at Risk of Flooding

Conservation Area C

Ancient Monument A

GO-NW Graphics Unit
© Crown Copyright 1995
Based on the Ordnance Survey Map

ECONOMIC DEVELOPMENT

The Introduction to the Management Plan has described how the Mersey Estuary became industrialised and urbanised with relatively little concern for its natural resources. The result was a degraded and polluted Estuary which was undervalued in terms of its potential and threatened by further deterioration. Changes in commercial navigation, in technology and in industry and commerce in the 1960s and 1970s brought about large elements of that deterioration. Significant stretches of dock and waterfront and canalside industrial areas became redundant; the Liverpool South Docks was perhaps the most obvious single example, but the position was similar elsewhere if not quite so obviously.

At this point local communities, the public authorities and industry could have turned their backs on a badly polluted river, lined with decaying buildings and infrastructure and left the Estuary to neglect. This was not to be, however, as two more positive factors recognised the Estuary as a source of potential strengths and opportunities. First, there has been a growing revaluation of the natural resources of the Estuary, its ecology, wildlife, its habitats and its environment. This factor has led to programmes to improve water quality and to protect and manage natural resources more responsibly.

The second factor has been a re-appraisal of the economic potential offered by the Estuary, which could once more be seen as an opportunity rather than a problem. This has resulted in new, more competitive positions for commercial navigation and the port sector, in the reclamation of derelict land and water space, and, in new riverside development including open space. At the same time there has been an increasing recognition that the Estuary has an important role to play in developing tourism in the wider sub-region.

The relationship between these two factors is not an easy one, however, and the commissioning of the Management Plan represents a recognition of this concern. An earlier imbalance had produced unsustainable development and current problems; more recent changes provide the occasion to bring the economy and the environment into a more sustainable relationship. This viewpoint emerged strongly from the consultations and specialist studies carried out in the early stages of preparing the Management Plan.

The recognition of such issues has informed the development of strategic objectives and policies for this economic development section of the Management Plan. It is important to stress that these three areas are themselves interrelated, and, that they are seen as set in the context of the Estuary's resources and its natural systems as a whole. They cover the following policy areas:
- Commercial Navigation and Port Development
- Urban Regeneration
- Tourism.

5 : Commercial Navigation and Port Development

The contribution of the Estuary's ports and port-related employment is of the first importance to the Merseyside and north-west Cheshire sub-region. The ports themselves with their docks, wharves and storage spaces provide the key elements, being the trans-shipment points between land and sea. In terms of the 'Atlantic Arc' European super-region, they constitute a 'Gateway' for world-wide, deepsea routes to connect with routes to northern, western and southern European ports as well as landside road and rail connections within Britain itself. The Liverpool Freeport (expanded in 1992 into Birkenhead Docks), Britain's most successful, and the Euro-Rail Terminal for Channel Tunnel freight illustrate the strength of the facilities being developed and their potential for the future.

The ports have two further principal roles: as a main port for Irish Sea ferry routes, and, as a point of access for port-related industries, most along the Manchester Ship Canal (e.g. Ellesmere Port, Stanlow, Runcorn and Trafford Park). The Estuary's smaller ports, wharves and canalside docks provide complementary facilities to those of the main operators. However, ports and their related employment are subject to continuing change, and their operation and development also have wider effects on estuaries, their natural and physical resources and on residents. The Mersey and its ports are no exception to this general rule.

The navigation issues which require attention in the Management Plan focus on three general areas. First, supporting the continued prosperity of the Estuary's ports and their related industrial and service employment; secondly, calling for attention to be given to concerns about the future dredging spoil deposits; and, thirdly, on working out agreements for more widely acceptable mechanisms and levels of evaluation for the treatment of environmental as well as economic interests in development decisions.

The three issues indicate the Management Plan's incorporation of the European Union's principle of sustainable mobility into its objectives and policies, providing for seaports and commercial navigation to pursue their economic roles while containing their harmful effects on the environment.

OBJECTIVE

To support the continued commercial and economic development of the Estuary's ports and port-related employment areas compatible with the Management Plan's environmental policies.

POLICIES

Maintaining Commercial Navigation Channels

It is essential to maintain access for commercial shipping to the Estuary's docks which, together with dockside storage and processing industries, and, their supporting services, provide the foundation of the Estuary Zone's economy. Continued access requires the dredging of navigation channels in the tidal Estuary and dredging for the maintenance of water depth in the Ship Canal and in the docks. These are costly activities which are subject to change from time to time, including changes in trading conditions and in the river's regime.

The influence of commercial shipping requirements on the Estuary (e.g. in terms of land reclamation, the alteration of the shoreline, the construction of training banks and the maintenance of navigation channels) has been discussed in the introduction to the Estuary Dynamics Management Plan Objective (page 19). Overall it is considered that the physical system of the Estuary has adjusted to these changes and that present activities allow it to function in a semi-natural and self-sustaining way compatible with wildlife concerns.

The future management of the Estuary requires that account is given to both economic and environmental considerations in maintaining commercial navigation channels. Policy NV1 thus focuses on the commercial role of the Mersey's port functions and the central issue of maintaining access from Liverpool Bay, while setting this requirement in the context of the Plan's environmental policies with the principle of sustainable mobility complementing that of sustainable development. The related question of the disposal of dredging spoil, some of which has its source in channel maintenance operations, is considered in Policy NV6 below.

NV1 *Subject to economic and environmental considerations, partner organisations should seek to maintain commercial navigation channels from Liverpool Bay to the Estuary's ports and Ship Canal and their associated docks.*

New Developments and Shipping Access

As discussed above, the central importance of the ports and port-related employment to the Estuary Zone's economy depends on continuing access for commercial shipping. Because of natural tidal conditions and sedimentation processes it is necessary to maintain dredged channels. It is also necessary to protect those existing channels and associated non-dredged navigation routes from developments (e.g. an extension of mining of marine aggregates) in the tidal section of the Estuary which could adversely affect the passage of commercial shipping.

NV2 *New activities or developments within the Estuary Zone should take into account the requirements for continued access by shipping using commercial navigation channels.*

Port Expansion Proposals

The Estuary Zone's ports operate in competitive markets with fast changing technologies which suggest a strengthening of the traditional role of the ports as interfaces for shipping, inland distribution (road and rail), storage and processing. Over the last quarter of a century large stretches of older docks and port-related employment areas have become disused and new activities introduced. Within existing port estates there have been expansion projects and the provision of new facilities (e.g. the PowerGen coal trans-shipment facility in the Gladstone Dock), the intensification of existing uses and local reorganisations.

Because of their significant wider implications for other Estuary stakeholders, it is important that partner organisations are alerted to possible longer-term port development proposals at an early stage, especially in relation to the expansion of existing or new bases for port operations. Although changes are taking place within existing port estates almost continuously, expansions beyond the estates only occur from time to time, but when operators decide to make such new investments they will want to move quickly.

Consideration thus needs to be given to identifying sites with long-term potential for port operation (e.g. factors such as depth of water/existing quayside walls/surface area) and adjacent areas of port-related employment potential. The likelihood of possible sites being needed for port development can then be evaluated against other uses in the existing developed parts of the Estuary Zone. Such uses, for example, residential, recreational, retailing and general business, may be more flexible than port operations in their locational requirements. Similar procedures can also be undertaken for possible port sites in the undeveloped coastal or off-shore sections of the Estuary Zone. Estuary resource and environmental concerns will be important factors in both instances, but

Stanlow Dock on the Manchester Ship Canal, Ellesmere Port

will have a special significance in the case of proposals involving potential locations on the undeveloped coast or off-shore.

Where a case can be made for the identification of sites for port expansion (beyond those already committed, such as the Port of Liverpool's reserve for expansion at Seaforth), it may be desirable to reserve such sites and to make allowance for appropriate interim uses which will not in themselves prejudice eventual port development.

NV3 *Proposals for the expansion of port facilities beyond port operators' existing estates designed to maintain the competitiveness of the Estuary's ports and take advantage of new market opportunities should be identified at an early stage for consultation.*

Management Measure
NV3.1 *Subject to the encouragement of interim uses which would not prejudice development, suitable sites could be reserved for future port use.*

Port-Related Employment Areas
It is important that public policies both protect and enhance the Estuary Zone's existing port-related employment base (e.g. marine engineering, storage, processing, distribution and support service users benefiting from locations close to docks and wharves and the movement of goods from ship to land and vice versa). For instance, provision could be made for the recognition of port-related employment areas in planning policies. Priority should be given to port-related users when undertaking the renewal of disused docks and dockside sites and buildings in port-related development areas.

Complementary measures include the establishment of co-ordinated policies to control development, to provide for local environmental upgrading, and, support for bids to grant regimes (e.g. the existing City Challenge areas in Wirral and Bootle) and new projects involving Objective 1 funds and/or the Single Regeneration Budget. The Objective 1 Single Programming Document has, for instance, indicated a number of sites in the Merseyside sector of the Estuary Zone. These draw on existing technical studies which would provide a useful starting point for a more extended baseline study.

There is a further need to take account of the future possible needs for development at a variety of scales e.g. from the regionally important 'Flagship' sites suggested by draft Regional Planning Guidance (April 1995), to other quite extensive port employment-related sites both within port operators' estates and elsewhere in Sefton, Wirral and Ellesmere Port through to smaller sites in the Runcorn Canalside area. Some existing proposed sites and port-related employment areas are shown on Map 2.

NV4 *Areas and sites should be identified within the existing developed sections of the Estuary Zone where preference should be given to port-related employment.*

Management Measures
NV4.1 *Undertaking a technical study of the future demand for, and, supply of an appropriate range of scale of industrial, storage and distribution sites for new expanding and relocating port-related industries and services and their transport requirements in the port-related employment areas and elsewhere in the Estuary Zone. Any proposals arising from the study should be subject to an environmental appraisal.*

NV4.2 *Defining existing port-related employment areas within which preference will be given to actions, including local environmental up-grading where appropriate, supporting the retention of existing and attracting new port-related users.*

NV4.3 *Encouraging the availability of a range of sizes of sites, of types of potential users and of development programming priorities including land reclamation and infrastructure provision within port-related employment areas.*

NV4.4 *Supporting the Port of Liverpool's role as a European Gateway for deepsea and shortsea traffic through providing for the development of port-related activities on a regional strategic 'Flagship' scale at a site adjacent to the port and rail freight terminal at Seaforth.*

NV4.5 *Providing, either individually or in partnership with other agencies, for the reclamation and reuse of appropriate disused port or dockland sites for small-, medium- and larger-scale port related users.*

Port-Related Business Enterprise

Management policies and measures should also seek to stimulate and support the industrial and commercial activities providing port operation and port-related employment. These actions may be relatively small-scale and specialised; they may relate to new activities and firms (e.g. in the offshore oil and gas support sector as more specifically considered in the Urban Regeneration section of the Management Plan) or they may cover well-established firms facing changing market conditions and opportunities.

Both existing and new firms need to be open to innovation and the application of new technologies in order to maintain the Estuary Zone's port competitiveness and to take up new opportunities. Given the firm-size structure of the port-related employment areas, there may be particular scope for recognising the needs of small- and medium-sized firms in the port-related employment sectors.

NV5 *Support should be offered to actions designed to promote the commercial and economic development of the Estuary Zone's port and port-related firms and locations.*

Management Measure
NV5.1 *Continuing and developing financial, business advice, and, training and enterprise support schemes related to port operations and port-related employment, including skill enhancement, starter firms, and, the development of small- and medium-sized firms.*

Dredging Obligations and Disposal

For the foreseeable future there will be continuing dredging obligations to maintain commercial navigation channels and water depth in the Estuary Zone's docks and in the Ship Canal and for other statutory (e.g. land drainage) reasons in the Ship Canal. Some dredging spoil is deposited at sea, while landward deposit grounds already cover extensive areas in parts of the Upper Estuary at Woolston and in the Frodsham-Helsby Marshes area.

Policy concerns focus on the chemically contaminated nature of the dredgings from the Ship Canal, from docks and the Estuary and the available capacity of existing landsite deposit grounds. A related concern is whether policy developments may require additional landsite deposits for estuary-derived dredgings, especially if there are limitations on disposal at sea, which would have adverse impacts on landscape and wildlife.

The scope for local initiatives on these issues, however, may be quite limited by the operation of international conventions (e.g. the Combined Oslo Paris Convention, including the 1991 Oslo Commission guidelines, and, the London Dumping Convention) and national policies. On land the proposed Environment Agency is set to take over existing responsibilities from predecessor bodies. What emerges is a need for the issues and policy conclusions to be more widely explored and clearly understood.

Although a good deal of detailed work is already being undertaken by the responsible agencies, the Management Plan has identified the scope for addressing the issues of dredging obligations and the subsequent disposal of the dredged material in a more coordinated and open fashion. This can be viewed as a technical assessment rather than a policy debate. The Ministry of Agriculture, Fisheries and Food, the port authorities, the Mersey Conservancy, the NRA, English Nature and the local authorities are well placed to lead such an exploration, drawing in community and specialist voluntary sector interests at appropriate stages.

NV6 *A technical assessment, which will need to include environmental protection and sedimentation considerations, should be made of future dredging obligations in the Estuary Zone and any disposal considerations. The assessment should cover dredging obligations for the maintenance of existing navigation channels and water depth in the Estuary's docks and the Ship Canal and for other statutory reasons.*

Management Measure

NV6.1 *The assessment, some items of which are already being undertaken, should be structured on a comprehensive basis to:*

(i)   establish the scope and direction of future policies, including international agreements, relating to the deposit of dredgings on landsites and at sea;

(ii)   define criteria for assessing proposals for additional or alternative deposit grounds or locations;

(iii)   specify the rate of current and likely future dredgings and their deposit requirements;

(iv)   establish the capacity and acceptability of current deposit grounds;

(v)   identify areas of search and alternative sites;

(vi)   incorporate environmental appraisals of any specific additional or alternative deposit ground proposals.

Permitted Port Development Rights: Code of Practice

One of the port and navigation issues addressed by the Management Plan is that of conflicts between commercial priorities, legal obligations and environmental management considerations. In terms of legal obligations, many port and harbour development and maintenance activities, for example, those by port operators on operational land for the purposes of shipping and in connection with the loading and unloading of goods and passengers, have been classed as permitted development and as such do not require the operation of the normal planning control system. Map 2 indicates the existing operational estates of port operators within the Estuary Zone where such permitted development rights may be exercised. Such developments may not only have impacts within the docks and harbours concerned, but may also have wider effects on neighbouring areas and other parts of the Estuary. The General Permitted Development Order 1995 sets out revisions to this position and schemes categorised as permitted development now require planning permission, to which conditions may be attached, if they fall within the list of developments for which an Environmental Assessment is needed. There is also a procedure for developers to ask for an opinion from the local authority on whether their proposals require an Environmental Assessment. Nonetheless, there is still concern that some nuisance-creating developments are left outside planning control.

The Management Plan approach does not regard direct external regulation as being the only way of dealing with the environmental challenges facing the Estuary Zone's port operators and port activities. Instead it suggests that the encouragement of processes of self-regulation to monitor, control and remedy activities harmful to the environment within a context of goals and objectives shared with the wider community of public and local authorities, residents and specialist interests can also provide a practical and advantageous way forward.

This would provide for the specific application of the general 'test practice' approach mentioned above in relation to policy PC1 where it is recommended that partner organisations should consider the adoption of a recognised environmental management system and appropriate code of practice and actively promote the benefits of such approaches within the Estuary zone.

The British Standards Institiute's BS7750 on Environmental Management Systems (1992); revised 1994) is indicated above as the recommended recognised procedure. The European Union's Eco-Management and Audit scheme (EMAS 1993), which is broadly comparable with the British Standard scheme, is a further example of interest. In terms of codes of practice relating to ports, the *Environmental Code of Practice* adopted by the European Sea Ports Organisation (1994) and the British Ports Federation/ Association's *Environmental Statement and Environmental Code of Practice* are illustrative of the range of topics and measures being considered generally.

Given the opportunity for improving environmental quality offered by the Management Plan, it is suggested that one focus might be provided by the development of a Mersey Estuary Code of Practice to complement the active encouragement given to the adoption of environmental management systems within specific agencies. This would imply the creation of routine opportunities for consultation between port operators and users and other partner organisations, and, joint working to establish shared objectives and procedures for an improved environmental quality. These could provide the basis for improving more general working relationships between the different interests in the Estuary Zone and for achieving economic and environmental benefit. The critical tasks for partner organisations in establishing such a Code of Practice will be reaching agreement to support regular exchanges of information and ideas, and, stimulating a general commitment to making the approach work effectively on a continuing basis.

NV7 *Partner organisations should promote the adoption of best practice to enable the Estuary's port operations to fulfil their commercial and economic roles while minimising harmful environmental effects.*

Management Measures

NV7.1 *Establishing and adopting an Environmental Code of Practice for the Mersey Estuary which will encourage port operators and users to improve environmental standards, establish environmental management systems, and, promote consultation with local communities, local authorities, environmental agencies and voluntary bodies.*

NV7.2 *Providing through the Environmental Code for the environmental assessment of, and consultation procedures about, significant proposals in the Estuary Zone which fall within existing permitted development rights and are exempt from development control. A useful first step would be to establish a comprehensive baseline of information about the nature and extent of port operators' existing permitted rights.*

6 : Urban Regeneration

The Mersey Basin Campaign's overall objective is to contribute to the regeneration of the region's environment and quality of life through waterside improvement as well as water quality initiatives. The reclamation and regeneration of derelict or disused land and buildings in the Estuary's bankside and waterfront areas can therefore be seen as a useful economic as well as environmental counterpart measure to the improvement of water quality. Central Government's consultation draft on Regional Planning Guidance (1995), the Merseyside Objective 1 Single Programming Document, and the policies and plans of the Estuary Zone's local authorities more generally place a strong emphasis on urban regeneration. The on-shore benefits of the oil and gas explorations and developments in Liverpool Bay also provide opportunities for maintaining and strengthening the economy of the immediate and the wider Estuary Zone.

As set out in the Land Use and Development objective and policies above, the Management Plan's proposals for economic regeneration focus on encouraging new developments within existing developed areas especially along the waterside and in adjacent regeneration areas and sites in the Estuary Zone. This focus extends to encouraging appropriate uses - for instance business, recreation, housing, tourism and heritage, including suitable mixed-use projects - to such locations, and, supporting improvement initiatives, including safeguarding the value-added of existing regeneration projects. The precise determination of appropriate uses on specific sites remains of course a matter for determination by local authorities through their Development Plan policies and proposals and development control decisions.

OBJECTIVE *To promote the regeneration of the Estuary Zone through maintaining and realising the distinctive potential of its existing developed waterfront and bankside locations and in adjoining areas.*

POLICIES

Stimulating Waterside Regeneration

The opportunities for 'Flagship' scale projects in the central Liverpool section of the Narrows have been proposed by Merseyside Development Corporation for the King's Waterfront site in Liverpool's South Docks, and, by Mersey Docks and Harbour Company in partnership with P & O Developments at Princes Dock. These exciting and prestigious sites have been marketed with the scope for major projects incorporating uses such as high quality offices, leisure facilities, some retailing, a hotel and residential apartments.

Other Estuary Zone 'Flagship' opportunities exist for the Cammell Lairds site in Birkenhead, Wirral, and, British Aerospace's proposals for the Liverpool North Airport (within the context of the partnership approach to the regeneration of the Speke-Garston area and the proposed development of Liverpool Airport). Mention has been made earlier in the Management Plan (Management Measure NV4.4) of support for the development of port-related activities at a regional strategic 'Flagship' scale at a site adjacent to the port and rail freight terminal at Seaforth.

Besides the 'Flagship' type of opportunities, the Estuary Zone also contains a range of different scales and types of development opportunity. Although some large-scale opportunities do exist for employment-related greenfield development on inland sites in and immediately adjacent to the Estuary Zone (e.g. at Manor Park East, Runcorn and in Ellesmere Port and at Bromborough), opportunities elsewhere are more constrained.

However, there are locally important opportunities for the reclamation of land from old industrial uses at Monks Hall and Arpley in Warrington for new light business uses, a variety of projects in New Brighton covering commercial, residential and leisure uses, and, new warehouse/light industrial uses on the site of an in-filled dock in Runcorn.

This diversity and potential suggests the need for a co-ordinated view of existing development opportunities throughout the developed part of the Estuary Zone and for the identification of further new locations (e.g. at Weston Point Dock). There is scope too for building on and extending the already extensive local experience of partnership arrangements between landowners, private and public development agencies, investors, potential occupants and building firms for regeneration projects. As much as the larger mixed-use schemes, there is also the potential for projects directed towards schemes designed for business support for the enterprises which form an important part of the Estuary Zone's industrial structure.

RG1 *Partner organisations should seek to identify regeneration opportunity areas and sites along the Estuary's existing developed waterside and elsewhere in the developed sections of the Estuary Zone.*

Management Measures

RG1.1 *Supporting the availability of an appropriate portfolio of regeneration proposals ranging from 'Flagship' schemes to more limited in-filling or rehabilitation and re-use projects.*

RG1.2 *Supporting the continued environmental upgrading of the Estuary Zone's existing developed areas (including new landscaping) and the modernisation of the infrastructure (including communications, site access, off-street parking and public safety concerns).*

Regeneration-Related Business Enterprise

As with the Zone's port-related employment areas, there is a need for management policies and measures to stimulate and support local industrial and commercial activities in regeneration areas. Tourism, recreation, cultural and media-related activities should be included within the scope of these policies and measures.

There is scope for learning in this and the Port-Related Business Enterprise policy area from the experience of the financial, business advice and training and enterprise support measures provided by Merseyside Development Corporation, and the Estuary Zone's Training and Enterprise Councils (TECs) and local authorities. Local and industrial sector business associations should also continue to be supported in their endeavours to improve business confidence and prospects and there may be opportunities for Chambers of Commerce to develop new partnership agency roles in this respect. With the ending of the Corporation's work reportedly in sight in the next three years, Management Plan Partner Organisations have an exciting opportunity to create new local institutional mechanisms to continue and expand essential business support service in a wider-Estuary Zone framework.

RG2 *Support should also be offered to actions designed to promote the commercial and economic regeneration of the Estuary Zone's existing developed areas.*

Management Measure

RG2.1 *Continuing and developing financial, business advice, and, training and enterprise schemes related to existing and future regeneration opportunities (e.g. in the tourism and recreation sectors) including skill enhancement, starter firms, small- and medium-sized firms.*

Reclaiming and Re-using Derelict Land and Buildings

The successful promotion of both waterside and port-related regeneration opportunities will also require supporting environmental upgrading (including the removal of dereliction, land treatment, new landscaping), infrastructure modernisation (including communications and access), and, related development/improvement and enterprise measures. The local and public authorities have established considerable experience in designing, funding and implementing such projects, often in partnership with the private sector.

The Management Plan seeks to bring these activities into a comprehensive and co-ordinated relationship with each other, so that individual projects are mutually supportive and can be seen as part of an overall design. The Mersey Basin Campaign's River Valley Initiatives (RVIs) such as those established for the River Alt (the Alt 2000 RVI) and the Weaver are good examples of overall, co-ordinated approaches established within financial and time programme frameworks. They enable diverse resources to be brought to bear in specific locations and offer a focus for local involvement for all sectors of the community. The institutional challenge will be to adapt the approach for the rather different context of the Estuary Zone.

In addition to new development and to redevelopment projects, there is a parallel need for measures to consolidate the position of existing employment areas in or adjacent to docks and waterfronts. A start has already been made in many areas to promote supportive commercial and industrial improvement policies. These involve the smaller-scale rehabilitation or conversion of some buildings, the assembly and servicing of sites for a few new projects, security improvement, provision for better on-street and off-street parking and for better vehicular access. As well as rehabilitation and conversion, appropriate new uses and projects, and, infrastructure modernisation, there is also scope for including greenspace landscape as part of an overall approach to waterside environmental upgrading.

RG3 *In the Estuary Zone's existing developed areas, and, in other areas as appropriate, partner organisations should promote coordinated programmes of remedial measures to reclaim and reuse derelict land and to restore and renew disused and derelict buildings.*

Management Measures

RG3.1 *Undertaking a coordinated survey of existing derelict and disused sites along the Estuary's waterside and elsewhere in the Estuary Zone and reviewing options for the application of remedial measures.*

RG3.2 *Establishing programmes for the coordinated and phased reclamation of identified sites and buildings and for their appropriate afteruse.*

Safeguarding Realised Improvements

The sustainable regeneration of the Estuary Zone's developed areas depends not only on future efforts, but also on securing and maintaining the quality of existing waterside areas where there has already been regeneration investment. These include examples such as Liverpool Pier Head, the Albert Dock and Birkenhead-Woodside (in the Narrows), more local and diverse land reclamation and improvement sites (often with a significant degree of environmental and open space quality) in Widnes and Runcorn, at the Ellesmere Port Boat Museum, and, the former International Garden Festival site (in the Inner and Upper Estuary sections). Other such areas include existing residential locations with distinctive heritage character (e.g. Rock Park, Wallasey-New Brighton, Cressington-Grassendale, and Waterloo) and sites in central Warrington.

For the most part what is required is a continuing management awareness of and concern for quality and standards and levels of funding. The particular actions required include: maintaining a safe and secure environment; prompt repairs and regular maintenance (including hard and soft landscaping as well as of buildings); and, routine cleansing and graffiti removal. Occasionally there may be opportunities for the enhancement of existing features, in-filling improvements and rehabilitation or smaller-scale new development projects linking hitherto un-connected areas. Not all activities presently located in or near such areas are necessarily compatible with their neighbours (e.g. in terms of traffic, odour, noise), and while these matters are dealt with more specifically in the Water and Pollution Control section of the Management Plan where specific local circumstances permit, then re-location (if environmentally acceptable elsewhere in the developed Estuary Zone) and/or ameliorative action (e.g. off-street parking, screening, modifying processes) should be promoted by partner organisations.

In some parts of the developed Estuary Zone, special investment funds or agencies (e.g. City Challenge and the Merseyside Development Corporation) have been provided for regeneration initiatives and the care of the built environment, but will in the near future be reaching the end of their operational support. In such cases there will be a need for consultations and agreement well in advance, for replacement mechanisms to be put in place by the local authorities and other partner organisations from their usual funding sources and for the continuing provision of services at maintained standards.

RG4 *In those areas of the Estuary Zone where regeneration is underway or has been effectively completed, continuing support should be offered for safeguarding realised improvements in environmental quality.*

Offshore Oil and Gas Resources

Exploration for oil and gas reserves in Liverpool Bay has proved successful in an area adjacent to the Estuary Zone's maritime boundaries. BHP (Hamilton Oil) has begun the installation of platforms and pipelines which will link the Hamilton (gas), Hamilton North (gas), Douglas (oil) and Lennox (oil and gas) offshore fields to the Douglas complex and control centre. From here a further pipeline will take gas to a terminal at Point of Ayr and thence to PowerGen's new power station at Connah's Quay; oil will be shipped out using an offshore tanker loading unit.

The beginning of production in the mid-1990s provides the opportunity for developing a 20+ year relationship with the Estuary Zone's industrial, commercial and transportation facilities. A supply base for an offshore support vessel has been established in West Hornby Dock, Bootle while helicopters based at Liverpool Airport take personnel to and from the construction, drilling and operation sites. Other localities, however, already have long-term promotional activities relating to the Irish Sea oil and gas region in place. In Merseyside a start has been made in this direction by business firms and some local authorities. There is scope, nonetheless, for a wider-based scheme involving the commercial sector, the chambers of commerce and the Training and Enterprise Councils (TECs) in a more comprehensive initiative.

The consideration of licences takes environmental concerns into account through the Department of Trade and Industry's consultation procedures. There would be local advantage, however, in partner organisations pooling information, at a North West regional level, and in coordinating their consultations for future bidding rounds through a wider area body such as the Irish Sea Forum.

RG5 *Subject to the protection of existing commercial navigation channels and the natural environment of the Estuary and its adjacent coastland, exploration for, and the development of, offshore oil and gas reserves in Liverpool Bay should be supported. Related on-shore activities and developments should also be supported where these are located in port operation and port-related employment areas and as appropriate elsewhere in the Estuary Zone's existing developed areas.*

Management Measures

RG5.1 *Promoting the industrial, service and commercial contributions to oil and gas production in Liverpool Bay on a long-term, multi-agency and Estuary-wide basis.*

RG5.2 *Supporting activities at a North West and Irish Sea level to examine the implications of oil and gas licensing, exploration and production, and to coordinate responses to licensing rounds and policy issues and proposals.*

Marine Aggregate Resources

Commercial dredging of marine aggregate for construction industry purposes is currently carried out at sites in Liverpool Bay and the Mersey Estuary. Licenses are issued by the Crown Estate Commissioners if the necessary 7-month Government View Procedure (involving extended consultation and study) reports favourably and if the Acting Conservator of the River Mersey does not withhold consent on Conservancy grounds in the Conservancy's area of jurisdiction. Subject to the protection of navigation and environmental concerns, this commercial marine aggregate dredging in the Estuary Zone should continue to be supported.

RG6 *Subject to the protection of existing commercial navigation channels, the natural environment of the Estuary and its adjacent coastland, coastal defence requirements, and, evidence of a satisfactory understanding of physical processes, support should be offered to commercial exploration for and winning of marine dredged aggregates in the Estuary Zone.*

7 : Tourism

Tourism has long influenced the development of the coast and, latterly, has come to play a vital role in the economic regeneration of Merseyside. In the Upper Estuary the tidal reaches are a less significant asset but both the River above the weirs and the canals provide attractive features.

Interest is, necessarily, focussed on the Narrows where the Pier Head buildings, the Albert Dock, the Ferries and the Birkenhead river frontage so clearly demonstrate the importance of the River in the history of Merseyside. These are the icons of Merseyside: their symbolic value in projecting the image of the place is incalculable; their care, their environment and their use are of the first importance to the tourist industry, but they do not function independently. Attractions at New Brighton, Ellesmere Port, Runcorn and Warrington also play their part (see Map 2). Links with the development of tourism facilities elsewhere on the shores of the Estuary and in the rest of the city are essential to the continuing success of recent initiatives and to the growth of the industry as a whole.

While tourism in the Estuary Zone is likely to be heavily orientated to the built up waterfront, the importance of the open shores must not be ignored. These offer a different range of opportunities for enjoyment and a contrasting environment. The conservation of the open shore will be an over-riding issue not only for the many reasons mentioned in the chapter on Estuary Resources, but also on the grounds of maintaining the widest possible range of tourist attractions.

OBJECTIVE *To realise the potential of the Estuary as a focus for tourism*

The Boat Museum, Ellesmere Port, develops the navigational theme

POLICIES

Promoting and Maintaining Diversity in Tourist Attractions

It will be important to attract more people to the riverside so as to encourage visitors to understand the many facets of the Estuary in an enjoyable way and to appreciate its dramatic landscape qualities. This will help to improve the image of the Mersey both among local people who, together with their friends and relations, form the majority of visitors, and among tourists from elsewhere in the UK and overseas. Indeed, the Estuary's long shoreline and maritime views are a great asset in developing the tourist industry; the attractions of the waterside for people enjoying their leisure are widely recognised.

New attractions have been introduced and established ones repackaged in recent years - the Battle of the Atlantic display and the Mersey Ferries are examples. Such developments, together with special events like the 'Return of the Tall Ships', have significantly increased the diversity of attractions and thereby built up custom. It will be essential for this process to continue with developments such as new leisure retailing venues of various kinds and new cultural venues and it is equally essential that the necessary support services in transport, communications, catering and accommodation develop at the same time. Only by creating greater diversity and making the tourist attractions as user friendly as possible will the potential of the Estuary to contribute to the growth of tourism in Merseyside and the North West be fully realised. At the same time, the development of the tourist industry will contribute to economic regeneration in the Estuary Zone.

The contrasting recreational opportunities on the undeveloped shores also allow for diversity since there a more natural environment can be enjoyed. Only low key facilities will be suitable on these shores; more formal tourist facilities will be better suited to the urban waterfront.

TR1 *Partner organisations should seek to build on the success of existing facilities and to promote and maintain diversity in the range of tourist attractions.*

Management Measure
TR1.1 *Identifying activities, facilities and sites which could be included in the portfolio of proposals promoting regeneration (RG1).*

Development of New Facilities for Tourists

Tourism and leisure facilities are included among the uses suited to the economic regeneration and the conservation of the waterfront.

One of the most important developments will be new accommodation. This will help secure a better share of the profitable markets in services for staying visitors, which are currently underdeveloped by comparison with those for day trippers and people visiting their friends and relations. Quality hotels are needed to cater for visitors on business and to support efforts to attract the conference trade to Merseyside. At the same time, more budget accommodation would enable more young people and groups to stay in the area. Other new leisure facilities too, for example, a conference centre or sports and entertainment facilities, may be considered.

However, it will be essential to ensure that, if riverside sites are chosen for these new developments, the buildings are sensitively designed to take full advantage of the site and to enhance the waterside scene, and that they justify the use of a riverside location by their relationship to other waterside and tourist uses and by allowing for public access to the waterside.

TR2 *New facilities should be promoted as opportunities arise through redevelopment along the built up shores and in the docks, especially favouring those which are likely to attract a mix of overseas and domestic visitors and which do not duplicate existing attractions without adding anything new.*

Management Measure

TR2.1 *Identifying and promoting suitable waterfront sites for tourist development, paying particular attention to finding venues for high quality hotels and for budget accommodation.*

Securing a Critical Mass of Tourist Services

Concentrating effort in the metropolitan core seems more likely than dispersing new investment widely to help to secure what the North West Tourist Board describes as the 'essential critical mass of services' which will generate greater tourist activity. However, although the tourist venues in this area include many of the most popular free and paying attractions on Merseyside, they are part of a wider set of facilities. User friendly communications between attractions in the metropolitan core and other city centre facilities on either side of the River will be crucial to the success of all.

TR3 *Partner organisations should seek to encourage investment in the vicinity of the Narrows in order to underpin the range of existing facilities in this area and to encourage new support services to develop.*

Enhancing the Role of the Mersey Ferries

The Ferries bring activity to the river scene as well as providing a vital and historic link across the Narrows (see Map 2). It is important that they continue to operate in this way. Commuters account for a small but regular proportion of passengers but the majority of trips are made by visitors, mainly from the North West, with a significant 40% making repeat visits. The Ferries play an important role in major events given the co-operation with land transport managers which has been achieved on these occasions. For instance, crossing the River by Ferry greatly added to the enjoyment of spectators of the Tall Ships, and the cruises initiated in 1994 to celebrate the centenary of the Manchester Ship Canal have proved to be successful. The Company has increased the range of activities, especially cruises, in recent years and thereby enhanced opportunities for people to see much of the Outer Estuary from the water. The Company are looking for opportunities to develop new tourist attractions upstream and downstream but these are likely to depend on acquiring new boats as well as on the market.

TR4 *Partner organisations should support the Mersey Ferries in enhancing their role both as a tourist attraction and as a means for people to cross the River.*

Development on the Open Shores

As noted above, it is important to maintain the distinction between leisure resources and activities along the more natural open shores and those in the built up areas. This distinction is consistent with other policies in the Management Plan which seek to protect the open shores from development.

The Mersey Ferry

The undeveloped shores offer opportunities for outdoor recreation on the beaches and the river banks which complement the tourist attractions of the built-up waterfront and add another dimension to the range of things visitors can do. Growth in the tourist sector may well lead to an increase in the numbers using the Estuary's open-air recreation facilities, especially among those visiting friends and relations, but the majority of users are likely to continue to be local people. The following sections on Informal and Water-based Recreation outline the Management Plan's proposals for recreation on the open coasts.

TR5 *In areas of open coast, any tourist development should be low key and compatible with environmental objectives and should promote opportunities consistent with recreation policies as well as tourism.*

Reinforcing Links Between Tourist Attractions

It is clear that many people combine visits to several attractions in the area. It is important to encourage them to continue to do so in order to maximise their appreciation of all the Estuary has to offer and to maximise the benefit to the local economy of their spending. Visitors from within Merseyside, as well as those staying in Chester and Southport, can visit a wide variety of attractions along the Estuary in numerous combinations.

It will be important to capitalise on these assets. It will also be important to support existing agencies and authorities in presenting and marketing the tourist attractions of the Estuary as a whole and in relating these to the wider, regional, package of facilities.

Tourist operators stress the importance of good quality facilities. Ways should be sought in which relevant authorities can be supported in efforts to improve the management of the environment and to provide high

quality services at tourist venues, especially access, information and catering.

TR6 *Partner organisations should seek to reinforce links between tourist attractions throughout the Estuary and beyond.*

Management Measure
TR6.1 *Encouraging the relevant agencies to promote the tourist attractions of the Estuary as a whole and supporting them in enhancing the range and quality of tourist services.*

The Value of Special Events

The Outer Estuary is a superb location for large scale events on the water such as the Tall Ships. Very large numbers of people are able to reach the coast by public and private transport, for instance at Pier Head, New Brighton and Crosby. Other events, such as those held at Ellesmere Port Boat Museum, also play an important part in promoting a more attractive view of the Estuary and its surroundings. A programme of events up to the year 2000 is currently being developed and should be supported.

TR7 *The value of special events, especially the large-scale occasions, in promoting Merseyside and bringing tourists to the area, is acknowledged. Partner organisations should support wholeheartedly the promotion of future special events.*

Liverpool's Albert Dock attracts six million visitors per year

Figure 5i: Economic Development: Summary of Policy Areas, Policies and Management Measures.

Strategic Policy Area	Strategic Policy		Management Measure	
5 NV: Commercial Navigation and Port Development	NV1:	*Maintaining Commercial Navigation Channels*		
	NV2:	*New Developments and Shipping Access*		
	NV3:	*Port Expansion Proposals*	NV3.1:	Reserving Sites for Future Port Use
	NV4:	*Port-related Employment Areas*	NV4.1:	Defining Port-related Employment Areas
			NV4.2:	Undertaking a Technical Study of Sites
			NV4.3:	Encouraging Variety in Sites, Users and Programme Priorities
			NV4.4:	Supporting the Port of Liverpool's European Gateway Role
			NV4.5:	Providing for the Reclamation and Reuse of Port Sites
	NV5:	*Port-related Business Enterprise*	NV.5.1	Continuing and Developing Training and Enterprise Support Schemes
	NV6:	*Dredging Obligations and Disposal*	NV6.1:	Assessing Dredging Obligations
	NV7:	*Permitted Port Development Rights*	NV7.1:	Establishing an Environmental Code of Practice
			NV7.2:	Providing for Environmental Assessment in Areas covered by Permitted Development Rights

Figure 5ii: Economic Development: Summary of Policy Areas, Policies and Management Measures.

Strategic Policy Area	Strategic Policy		Management Measure	
5 RG: Urban Regeneration	RG1:	*Stimulating Waterfront Regeneration*	RG1.1:	Supporting Availability of a Portfolio of Regeneration Proposals
	RG2:	*Regeneration -related Business*	RG1.2:	Supporting Environmental Upgrading of Existing Developed Areas
			RG2.1:	Continuing and Developing Training and Enterprise Support Schemes in Existing Areas
	RG3:	*Reclaiming and Re-using Derelict Land and Buildings*	RG3.1:	Undertaking a Survey of Sites and Reviewing Options for Remedial Action
			RG3.2:	Establishing Reclamation Programmes
	RG4:	*Safeguarding Realised Improvements*		
	RG5:	*Offshore Oil and Gas Resources*	RG5.1:	Promoting Industrial, Service and Commercial Contribution to Oil and Gas Production
			RG5.2	Supporting Wider Activities to Examine Implications of Oil and Gas Licensing, Exploration and Production
	RG6:	*Marine Aggregate Resources*		

Figure 5iii: Economic Development: Summary of Policy Areas, Policies and Management Measures.

Strategic Policy Area	Strategic Policy		Management Measure	
7 TR: Tourism	**TR1:**	*Promoting and Maintaining Diversity in Tourist Attractions*	**TR1.1:**	Identifying Activities, Facilities and Sites to Promote Economic Regeneration
	TR2:	*Development of New Facilities for Tourists*	**TR2.1:**	Identifying and Promoting Sites for Tourist Accommodation
	TR3:	*Securing a Critical Mass of Tourist Services*		
	TR4:	*Enhancing the Role of the Mersey Ferries*		
	TR5:	*Development on the Open Shores*		
	TR6:	*Reinforcing Links Between Tourist Attractions*	**TR6.1:**	Encouraging Relevant Agencies to Promote Tourist Attractions of the Estuary
	TR7:	*The Value of Special Events*		

OTHER RELEVANT MATERIAL

The following material produced by the Study Team is also relevant to this chapter:

TOPIC REPORTS

1   Initial Consultation with Voluntary Organisations and the Private Sector

2   Navigation, Tidal Regime and Level of Use

6   Landownership and Tenure

9   Tourism

10 Emergency Planning

11 Fishing and the Mersey Estuary

OVERVIEW REPORTS

Mersey Estuary Management Plan : First Year Report

Mersey Estuary Management Plan : Report on the Consultation Exercise

MAP 2 ECONOMIC DEVELOPMENT

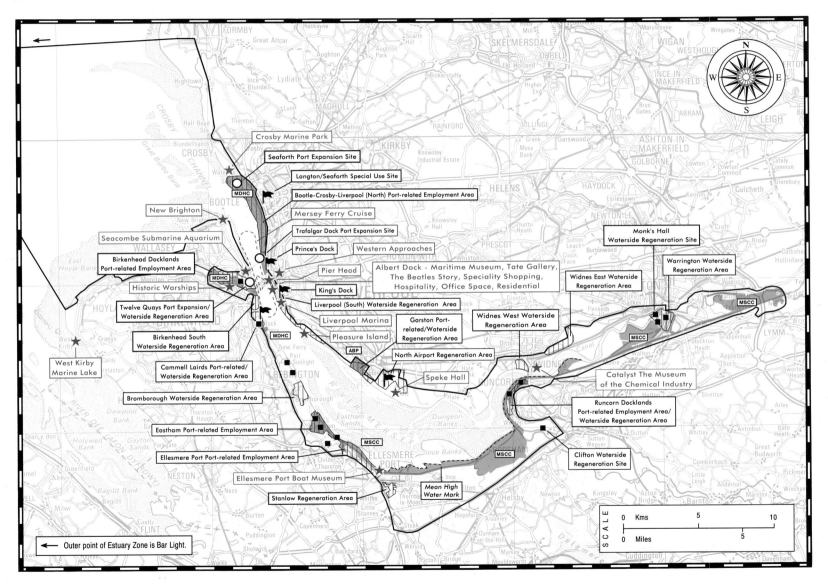

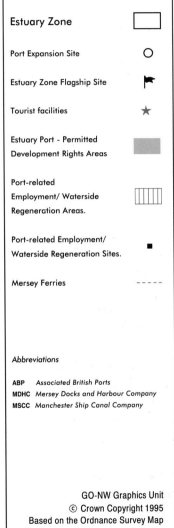

Estuary Zone

Port Expansion Site

Estuary Zone Flagship Site

Tourist facilities

Estuary Port - Permitted
Development Rights Areas

Port-related
Employment/ Waterside
Regeneration Areas.

Port-related Employment/
Waterside Regeneration Sites.

Mersey Ferries

Abbreviations

ABP *Associated British Ports*
MDHC *Mersey Docks and Harbour Company*
MSCC *Manchester Ship Canal Company*

GO-NW Graphics Unit
© Crown Copyright 1995
Based on the Ordnance Survey Map

Map labels:

Crosby Marine Park
Seaforth Port Expansion Site
Langton/Seaforth Special Use Site
Bootle-Crosby-Liverpool (North) Port-related Employment Area
Mersey Ferry Cruise
Trafalgar Dock Port Expansion Site
Prince's Dock
Western Approaches
New Brighton
Seacombe Submarine Aquarium
Birkenhead Docklands Port-related Employment Area
Historic Warships
Twelve Quays Port Expansion/ Waterside Regeneration Area
Birkenhead South Waterside Regeneration Area
West Kirby Marine Lake
Cammell Lairds Port-related/ Waterside Regeneration Area
Bromborough Waterside Regeneration Area
Eastham Port-related Employment Area
Ellesmere Port Port-related Employment Area
Ellesmere Port Boat Museum
Stanlow Regeneration Area
Pier Head
Albert Dock - Maritime Museum, Tate Gallery, The Beatles Story, Speciality Shopping, Hospitality, Office Space, Residential
King's Dock
Liverpool (South) Waterside Regeneration Area
Liverpool Marina
Pleasure Island
Garston Port-related/Waterside Regeneration Area
North Airport Regeneration Area
Speke Hall
Monk's Hall Waterside Regeneration Site
Warrington Waterside Regeneration Area
Widnes East Waterside Regeneration Area
Widnes West Waterside Regeneration Area
Catalyst The Museum of the Chemical Industry
Runcorn Docklands Port-related Employment Area/ Waterside Regeneration Area
Clifton Waterside Regeneration Site
Mean High Water Mark

← Outer point of Estuary Zone is Bar Light.

SCALE
0   Kms   5   10
0   Miles   5

RECREATION

The wide shoreline landscapes along the Mersey Estuary offer sharp contrasts between open coasts with sandy beaches and river banks typical of the countryside on the one hand, and the landscapes of dockland and industry, town and city centre on the other. The Outer and Inner Estuary shores allow unrivalled opportunities to enjoy fresh air and wide horizons; wildlife, especially over-wintering birds in great numbers; the urban scene and an historic waterfront. In the Upper Estuary, the gentle landscapes characteristic of the lower reaches of a major river can be enjoyed from a network of paths along the River banks and the historic canals.

Most forms of water-based sport and recreation can be practised somewhere on the Mersey or on the waters associated with it, such as its tributaries and the docks and canals built in the course of industrial development. The configuration of the Estuary in relation to the wind and the strong tidal currents make it a dangerous environment for the inexperienced or novice sailor, so that the Mersey is unlikely ever to become a major venue for boating. Nevertheless, the Estuary is the nearest extensive waterspace to the population of Merseyside and many others living in the Mersey Belt. It presents opportunities, which are valued by many sports enthusiasts, for all forms of boating, angling and wildfowling; for training; and as a base from which to explore other venues, for example round the Irish Sea. In the latter context the Mersey Estuary plays an important role because it is one of the few harbours of refuge on the west coast between Scotland and the Menai Straits.

This chapter addresses the issue of how to make full use of opportunities for recreation in the Estuary Zone wherever that is consistent with the other objectives of the Management Plan. It is important, however, to bear in mind that, while the Estuary offers unique opportunities, it is part of a wider scene. Resources along the shores are limited and will need to be used in the most appropriate ways and links to other venues and similar facilities in the open countryside and in the towns and cities will be vital.

Two strategic policy areas have been specified in relation to recreation, namely:

- Informal and Shore-based Recreation
- Water-based Sport and Recreation

8 : Informal and Shore-based Recreation

People use their leisure time in innumerable ways; the recreational uses considered here are those which are generally recognised as the more popular, in the sense that they are more frequently indulged, and which are, at the same time, the most likely to be compatible with the overall objectives of the Management Plan and are likely to benefit from a coastal or waterside location.

Walking is acknowledged to be the most popular recreational activity outside the home. While some will enjoy long walks, the majority go for shorter walks, often starting from or near home, and spending only part of the day, perhaps only half an hour, on their trip. Such trips often take in other ad hoc and somewhat passive activities, such as picnicking, playing with children or animals and pursuing an interest in natural or man-made surroundings. These activities take priority in considering policy. But that is not to underrate the importance of many more active pastimes which also rely on access to the waterside: that is, public transport and roads to the shore, paths along it with links to the hinterland, and back-up facilities such as toilets, car parks, information and refreshments. Examples of these more active pursuits include cycling, horse riding, orienteering and serious bird-watching. Fishing is in a similar category (there is an overlap with water-based recreation in this respect); motorbike scrambling is not included as it is most unlikely to be compatible with other objectives in the Plan, especially the protection of Estuary Resources. Sporting activities of the kind mentioned above are generally practised on an informal basis rather than in an organised and competitive way which would require space to be specifically laid out and solely dedicated to a particular activity. Many of them also appeal to individuals practicing on their own on a casual basis. For these reasons this section deals with 'informal sport and recreation' and concentrates on issues to do with access.

It is important to establish that terms used in the Plan such as 'routes' and 'paths' are used in a generic sense rather than in the legalistic sense of 'public footpath' or 'bridleway'. The plan addresses the general problem of access. In any particular location circumstances will determine how far different types of user can be accommodated. Ideally, there will be safe routes for walkers as well as cyclists, horse riders and, possibly, others. Where, for reasons of lack of space or safety, use has to be restricted, decisions will need to be taken as to how the route is designed and for whom it is designated.

Latterly, people have been encouraged to enjoy the shores of the Estuary and to use them for informal recreation. New points of access, new routes and country parks have been opened up. The Management Plan should continue to promote access wherever it is consistent with other objectives in the Plan and with public safety.

OBJECTIVE *To maintain, enhance and, where appropriate, extend public access to the shores of the Estuary so that people may enjoy informal sport and recreation in safety.*

POLICIES

Access on the Open Coast

The coast of the Outer Estuary is, traditionally, the most popular venue for informal recreation and is still the most easily reached by both public and private transport. Most of the coast and much of the open shore on the northern side of the Inner Estuary is accessible to the public. Local authorities have used opportunities as they have arisen to create new coast and riverside walkways and country parks and to establish long distance walks along the shore such as the Mersey Way (part of which

Bird-watching on Frodsham Marsh

doubles as the Trans-Pennine Trail), the Sefton Coast Path, and the Wirral Coast Path (see Map 3). Some missing links in these routes still need to be completed.

IR1 *In areas of open coast, the partner organisations should support the Countryside Commission's policy of ensuring that the entire Statutory Rights of Way network should be legally defined, properly maintained and effectively promoted by the year 2000.*

This is consistent with the policies adopted by several local planning authorities in the study area and with that stated in the Mersey Forest Plan (Mersey Forest Plan, 1993 - Policy R1).

Access on the Built-up Coast

In the built up areas, Pier Head and the Albert Dock in Liverpool offer the greatest concentration of opportunities for sport and recreation, both informal and commercial (see Map 3). Other facilities are associated with traditional riverside access points such as the ferry terminals and bridging points. New Brighton is unique in being the only venue that was developed as a seaside resort.

IR2 *In regard to access to the water's edge and river walls on the built-up coast and shores, public rights of way should be defined, maintained and promoted as they are on the open coasts whether they occur as streets, marine parades, promenades, public open space or in some other form.*

Such a policy will help to achieve consistency in accessibility and the management of routes along the shores of the Estuary.

It may be difficult, however, to retain some long-established access routes on the riverside in the wake of new development or operational requirements, for example, in the docks. These routes are especially important to particular members of the public, such as fishermen. Bankside operators should be encouraged to facilitate access, if necessary under licence, wherever possible.

Management Measure

IR2.1 *Seeking to maintain access to those parts of the shore and river walls which have traditionally been available to the public if new development or new operational practices on the bankside threaten discontinuance.*

Public Access and Waterside Developments

Taking advantage of the Estuary's waterfront for urban regeneration means providing access to it so that its facilities, its view and sights in the changing moods of light and dark and the seasons can be seen and experienced by the public. This measure is proposed in support of policy WR2, so that when new development projects are proposed the developers and planning authorities can include public access in their initial ideas. Planning authorities may wish to provide more detailed guidance in local plans and design briefs, especially where proposals covering links in more extended footpath systems are being considered e.g. the Trans-Pennine Trail.

IR3 *In developing waterside projects, appropriate public access to the banksides and the waterfront should be provided, especially where this will form a link in a more extended continuous path.*

Extending the Network of Paths

While local authorities place considerable emphasis on improving access for informal sport and recreation and stress the advantages of long distance routes, little information has been gathered on a systematic basis about the use people make of these routes for long or short walks. Along parts of the shore there are few paths connecting to inland networks or open spaces and, therefore, little opportunity to vary routes. This lack of variety may discourage the public from taking full advantage of the riverside paths.

Nevertheless, established initiatives which provide long distance routes have proved popular with walkers, cyclists and fishermen - in the built up areas as well as on the open coasts. Not all these routes are continuous nor are they all maintained to the highest standards throughout their length. Consultees have stressed the importance of the Mersey Way and the Trans-Pennine Trail. These two routes which share the same track along parts of the Estuary shore, illustrate the benefit of having an appropriately designed and designated route, which in this case serves walkers and cyclists at least, and the need to complete the links between isolated sections and provide and maintain routes to a consistent standard along their length. Efforts to do so throughout the Estuary Zone should be supported.

In carrying out this work, however, authorities will need to continue to be sensitive to the conflicts which can arise with recreational use. In some locations there is the danger of erosion of delicate habitats and geological features such as dunes, clay cliffs or salt marshes and disturbance of protected species, notably birds. Elsewhere, other interests, such as farming, will need to be taken into account.

IR4 *Advantage should be taken of any opportunities which offer to extend the network of paths leading to and along the shores of the Estuary and to enhance the opportunities for informal sport and recreation, provided that the increase in public access will be compatible with other objectives within the Management Plan.*

Management Measures

IR4.1 *Making every effort to achieve the appropriate design and designation of routes so that, in the light of local circumstances, safe routes can be provided for different modes of recreational traffic.*

The area covered by the Mersey Forest Plan overlaps with that of the Estuary Management Plan, significantly in the Upper Estuary. With regard to extending access and promoting opportunities for recreation, the objectives and policies of the two Plans are similar. Coordination in the refinement of policy and the development of specific schemes promoting access, sport and recreation will be important.

IR4.2 *Where the Estuary Management Plan and the Mersey Forest Plan overlap, particular attention should be paid to coordinating proposals to improve public access and informal sport and recreation so as to ensure that the proposals are compatible with the objectives of both Plans.*

Public Safety

Access to the waterside can be dangerous in certain conditions and some parts of the shore are unstable. Casual use of the water, for swimming for example, can seldom be encouraged. The design and management of

waterside routes must take these factors into account. In places it may be difficult to keep the path close to the shore but a route along the shore at low tide where the going is firm should be seen as an alternative not as a substitute, for a coastal path. IR4.1 refers to the need to separate pedestrians from vehicular traffic and to separate different types of recreational traffic - pedestrians, cyclists, riders.

Maritime and commercial operations, some involving hazardous substances or operations, require security. Public access is likely to be incompatible in some cases or to need very careful management.

There is also the question of personal safety; routes from which there is no escape by means of connecting networks can seem dangerous and may be so in the case of an accident.

IR5 *Public safety must be a prime consideration in the promotion and design of access for informal recreation on the shores of the Estuary.*

Some considerable sections of the shores of the Inner and Upper Estuary are inaccessible to the public. The reasons for this vary but the most common are the operational requirements of land uses on the banks, public safety, the constraints imposed by landowners, flood risk and the need to safeguard wildlife. The southern shores of the Inner Estuary, where the Manchester Ship Canal imposes a barrier, and parts of the River banks between Runcorn and Warrington, are among the least accessible sections and are likely to remain so (see Map 3).

Management Measure

IR5.1 *Accepting that public access should be limited in some areas for reasons of environmental protection and public safety.*

Encouraging the Use of Paths and Open Spaces

If people are to be encouraged to enjoy the shores of the Estuary they need good infrastructure as well as the network of appropriate routes mentioned above. This means public transport to, and car parks at, starting points for walks and attractions in both the open and built-up areas, such as country parks or places where refreshments and entertainment are available. Basic services such as toilets, and facilities which can be used by disabled people, the elderly and those with children are needed at the most accessible and popular locations. Helpful signposting is needed throughout.

The benefits of such infrastructure for informal recreation are widely accepted and are demonstrated by the popularity of some parts of the coast such as Formby, New Brighton and Spike Island. There has, so far, been no occasion to assess the accessibility, infrastructure and attractiveness of the Estuary as a whole but such an assessment would be valuable in highlighting deficiencies and pointing out places where improvements might best be concentrated so as to encourage people to enjoy the Estuary without endangering its other assets.

IR6 *The public should be encouraged to use the coastal paths and riverside walkways both in open and built-up areas by improving their accessibility, their design and the facilities which make them user-friendly.*

Extending Interpretation Services

An understanding of the variety of scenery available, the natural environment of the Estuary and of its role in the economic and social development of the surrounding towns should encourage people to visit the shores more frequently, to enjoy the experience in new ways and, above all, to appreciate the quality and importance of an unique asset. This last aspect is also addressed in the chapter on Implementation (see UM4). Many public sector and voluntary organisations are involved in interpretation and education and they have already achieved a great deal. Nevertheless, there are several places on both the open and urban shores where new and exciting interpretative facilities could be installed if resources were available (see Map 3).

IR7 *The services, in both the public and voluntary sector, which now provide interpretation and promote understanding of the natural environment and of the history of the River and the Estuary, should be encouraged to extend their activities both geographically and in terms of the subjects covered. They should be supported in making use of the latest technology. Initiatives to provide new services should be encouraged similarly.*

Several authorities provide comprehensive Ranger Services and interpretive facilities which show how much can be done to encourage the public to use and enjoy the shoreline. Not all the Estuary is covered to the high standards of the best of these services. The extension of these services and further exchange of information and co-ordination in practices and programmes of activities should encourage public response.

Management Measure

IR7.1 *Seeking to establish Ranger Services throughout the Estuary covering both the undeveloped and built-up shores to a consistently high standard.*

Maintaining Opportunities for Quiet Recreation

Noise pollution, especially on the open coast, is likely to be unacceptable to other users as well as to nearby residents and may disturb wildlife. Where vehicles are the cause there is a high probability of pollution and of erosion of soft surfaces on open coasts such as dunes and clay cliffs. While it is recognised that locations are needed for noisy sports, for example motor sports, no obvious sites present themselves on the shores of the Estuary where the risk of disturbance and erosion could both be reduced to acceptable levels.

IR8 *The plan should seek to maintain opportunities for quiet recreation, both active and passive, and, especially on the open shores, should seek to promote only quiet pursuits.*

9 : Water-based Sport and Recreation

Those who use the Outer Estuary or go to sea, such as keel boat sailors, sea anglers and speed boat enthusiasts, want access at all stages of the tide or, at least, where there is a long tidal 'window'. On the other hand, for some sports, such as rowing, access to sheltered waters with a constant depth is essential. However, several sports such as canoeing, fishing, dinghy sailing and water skiing can be adapted to either situation. As the water becomes cleaner, more people are likely to be attracted to it both for sport and casual recreation. The sports which seem most likely to grow in popularity are angling, motorised water sports, canoeing and windsurfing.

As is the case with the long-established water sports, these are all activities in which people can benefit from membership of a club where facilities, training and competition are available, but they may also be practised by individuals on a casual basis. In the Mersey Estuary, where many considerations such as safety, the needs of other users and environmental protection must be taken into account, the Management Plan must take careful cognisance of the potential for water-based recreation and the nature of participation. Clubs will have an extremely important role to play in promoting the safe and sustainable use of the water for recreation.

OBJECTIVE *To protect existing water-based recreation facilities and promote the appropriate development of new opportunities on the Estuary.*

POLICIES

Retaining Existing Facilities

There are only limited opportunities to moor and launch boats without the benefit of a lock and although they are concentrated in the Outer Estuary and above the Narrows, not all of them are available at all stages of the tide. Thus any loss of access, especially in the lower reaches of the Estuary, would seriously diminish opportunities for water-based recreation. In the Upper Estuary users rely largely on the waterway network made up of the various canals with, in places, links to the River (see Map 3). The lower reaches of the River Weaver offer a range of opportunities in the Inner Estuary Zone where opportunities on the Mersey are constrained for a number of reasons, not the least of which is the extensive exposure of mud and sand at low tide. Some enclosed bodies of water are also much used, particularly parts of the dock systems.

WR1 *Every effort should be made to maintain and enhance existing opportunities to get boats on to the water whether these are under private or public control.*

In some places, changes in the dynamics of the Estuary are reducing the depth of water to the point where sailing is restricted and could, if the process continues, become unviable. Problems on the River Alt and on the River Mersey in the vicinity of Rock Ferry and above Runcorn Bridge have been highlighted in consultation. Access needs monitoring and an understanding of the causes of deposition and the possible consequences of altering management regimes or redesigning facilities will be essential in planning to redress problems on the River Alt and in the Upper Estuary.

Sailing in Liverpool South Docks

Management Measures

WR1.1 *Undertaking studies to monitor access for recreational sailing where deposition is having an adverse effect and, where this is the case, determining whether adjustments in the design of facilities, the management of water regimes or dredging would enable depths of water suitable for sailing to be maintained while meeting other requirements of the Plan, particularly those concerned with Estuary dynamics.*

If it proves impracticable to ameliorate problems of deposition so as to retain sufficient depth for water-based recreation where it is currently available, the question of relocating facilities would arise. This, clearly, would involve as careful an assessment of the options as any other development.

WR1.2 *Recognising the need to relocate a facility for water-based sport and recreation if it is seriously and irreparably losing physical capacity and helping to find a suitable alternative site.*

There are very few public launching sites on the Estuary. It is particularly important for canoeists and others who are able to travel to a variety of waters, that parking and well-maintained changing facilities should be available to them when they use public launching sites. This issue is considered further in WR4.

Should established access points to the River, whether in public or private ownership, be affected by development proposals, it will be important to retain access. For example, the relationship between Liverpool Airport and Liverpool Sailing Club will, no doubt, be a consideration in development proposals for the former.

WR1.3 *Expecting developers to maintain and enhance existing access to the water and ensuring that associated landward facilities are retained.*

Retaining Access to the River Walls

This subject has already been covered in the section on Shore-based Recreation but is referred to again here because of its importance to fishermen. Fishing is a growing interest and cleaner waters will, in due course, enhance its attractions.

Existing routes along the river walls are greatly appreciated because they allow fishing to take place over the full range of the tide. The security of adjoining premises and the safety of individuals are legitimate concerns to authorities. Nevertheless, some arrangements, such as licensing, which allow for access should be practicable. Where the development or the management of riverside activities affects access, every effort should be made to retain it.

WR2 *Partner organisations should ensure that existing access routes along the River walls and the Estuary shores are retained and made available for recreational purposes, including fishing.*

Maintaining Existing Opportunities for Wildfowling

Wildfowling, as it is currently practised by responsible bodies and clubs under license, is consistent with the sustainable management of the marshes and of bird populations in the Inner and Upper Estuary. Its continuance will depend somewhat on maintaining the relative isolation of those marshes. Continuing co-operation between all the relevant authorities, and representatives of wildfowling and bird watching interests will also be important.

Consultees have expressed concern over illegitimate shooting on the northern shores of the Inner Estuary and the need for more effective policing, and possibly for shooting free refuges, in the vicinity of Oglet Bay.

WR3 *Existing opportunities for wildfowling should be maintained provided that this is compatible with the conservation of the natural environmental resources of the Estuary.*

Management Measure

WR3.1 *Encouraging the responsible authorities to monitor the degree, and the consequences, of illegal shooting on the northern shores of the Inner Estuary with a view to controlling it.*

New Opportunities and Facilities for Watersports

All forms of boating and sea angling would benefit from better facilities for access to the water as near to the sea as possible and at all stages of the tide. Canoeists are particularly anxious to obtain easier access to the full range of water bodies and to see the development of white water and slalom courses.

Opportunities for new developments are most likely to arise in the course of redevelopment in the urban area, especially of the older dock systems. Anglers will benefit from increased access to the shores of the Estuary which are consistent with access for informal recreation.

Other opportunities may occur with redevelopment, or new development, in the River; for example, the rebuilding of a weir or the construction of a barrage might allow for the development of a white water facility.

WR4 *The development of new opportunities and facilities for water-based sport and recreation should be encouraged where they will be compatible with other objectives in the Management Plan.*

Public launching facilities are limited throughout the Estuary and there are none on the Liverpool shore. Any increase in the capacity of existing facilities, such as those at New Brighton, or a site for a new facility, for example north of the Seaforth Docks, would require adequate landward facilities such as car and trailer parking and toilets and would raise the question of risk to commercial traffic because of the proximity of commercial shipping lanes. The cost of providing, maintaining and managing such launching sites would be considerable. Any development would need to be compatible with other objectives of the Plan, notably in respect of impacts on the dynamics of deposition or erosion. Nevertheless, better facilities would undoubtedly be welcomed, especially if they were near the sea, by sea anglers, canoeists and power boat enthusiasts.

Management Measure

WR4.1 *Setting up studies to establish the extent of demand for public launching facilities and the ways in which such a demand, if established, could be met. New or improved facilities would need to meet other objectives of the Plan.*

Sheltered Waters and Watersports Uses

Sheltered waters, such as non-commercial docks, marine lakes and canals, where water sports training and other water-based recreation can take place, which are suited to sports which need still water and where boats can be kept safely are an essential complement to the open waters of the Estuary and their potential should be exploited. In redeveloping the docks, for example, there may be an opportunity to extend training facilities or to incorporate another marina if it is possible to open a dedicated lock into the River. Liverpool Marina, MerseySports new premises in the South Liverpool Docks and the canal basins at Ellesmere Port Boat Museum demonstrate the potential of these waters for water sport and recreation and the visual attraction of activity on them.

WR5 *The incorporation in development proposals of water-based sport and recreation and training facilities on sheltered waters should be encouraged as an essential complement to the opportunities on the open waters of the Estuary wherever this is compatible with other objectives of the Plan.*

Public Safety

Many issues concerning public safety on the water, such as testing the competence of crews or the condition and equipment of craft and requiring insurance cover, are matters for national debate and well beyond the competence of the Management Plan. In the Estuary at present, the organisations responsible for safety, such as the Coastguards, local authorities, RNLI and the sports clubs, achieve a consensus on the measures needed to ensure safe practice and provide rescue services. This is done largely through voluntary co-operation in organising events, providing training courses and the imposition of club rules. There is, however, some concern that irresponsible behaviour by some casual boaters is creating dangers for others and that increased casual use of the water, especially by speed boats, will increase the risk of accident.

It will be important to continue to develop ideas on promoting safe practices, including codes of practice, especially where a sport attracts individual or casual users.

WR6 *Partner organisations should encourage the further development of management techniques and codes of practice which promote public safety on the water.*

The use of motorised craft, whether power boats or personalised craft, is likely to increase. The speed of these craft and their numbers already cause some concern, for example in the vicinity of New Brighton (see Map 3). It will be important to monitor both the numbers and the distribution of craft so that, if increased traffic should cause greater concern, measures to mitigate the problems could be discussed.

Lock Gate, Fiddlers Ferry

Management Measure

WR6.1 *Monitoring the numbers and distribution of leisure craft, particularly powered craft, using the Estuary.*

Resolving Conflicts Between Water-based Recreation and Other Users

The levels of recreational use of the waters of the Mersey Estuary seem, at present, to give rise to few conflicts either between recreational users themselves (aside from those relating to speed mentioned above) or between them and other users. Congestion occurs mainly at the few locations where access onto the water is available to several different groups of users and on those occasions when good weather and high tide coincide with holiday times and weekends. The principal exception to this is in the lowest reach of the River Weaver where water space is limited.

Much potential conflict in competition for water space is avoided through the mediation of sports clubs and their representative bodies working with the various authorities which control traffic on the River Weaver and access to the shore or through the medium of consultative groups meeting regularly to apportion time where space is limited, as, for instance, in the docks. The configuration of the Estuary and its tidal regime are helpful in protecting bird populations since most of the best areas are inaccessible from the water at most states of the tide. Thus the combination of geography and current systems for managing the recreational use of the various water bodies achieve a level of recreational use which is compatible with the objectives of the Management Plan.

In the future, however, changes could occur which could give rise to concern either about congestion in recreational and other uses of the water or about disruption to the natural resources of the Estuary. For example, increases in the numbers of craft might lead to unacceptable levels of congestion, erosion or to boats penetrating previously undisturbed parts of the Estuary. In such circumstances new measures would be needed to resolve conflict. Several options may present themselves, ranging from new management techniques, and the greater use of consultative groups to new regulations or full scale zoning schemes, all of which would have to be capable of being implemented and, in the case of regulations of zoning schemes, policed.

WR7 *Further to WR4, partner organisations should monitor the general levels of recreational use of the water in order to ensure that measures designed to reduce conflict between recreational users themselves and between them and other interests are functioning satisfactorily.*

Figure 6: Recreation: Summary of Policy Areas, Policies and Management Measures.

Strategic Policy Area	Strategic Policy		Management Measure	
8 IR: Informal and Shore-based Recreation	**IR1:**	*Access on the Open Coast*		
	IR2:	*Access on the Built-up Coast*	**IR2.1:**	Maintaining Access to the Coast
	IR3:	*Public Access and Waterside Developments*		
	IR4:	*Extending the Network of Paths*	**IR4.1:** **IR4.2:**	Appropriate Design and Designation of Routes Coordinating MEMP and Mersey Forest Plan
	IR5:	*Public Safety*	**IR5.1:**	Restrictions to Public Access
	IR6:	*Encouraging the Use of Paths and Open Spaces*		
	IR7:	*Extending Interpretation Services*	**IR7.1:**	Extending Coverage of Rangers
9 WR: Water-based Sport and Recreation	**WR1:**	*Retaining Existing Facilities*	**WR1.1:** **WR1.2** **WR1.3:**	Monitoring Access for Recreation Sailing Relocating Facilities Maintaining and Enhancing Access and Facilities
	WR2:	*Retaining Access to the River Walls*		
	WR3:	*Maintaining Existing Opportunities for Wildfowling*	**WR3.1:** **WR3.1:**	Monitoring Illegal Shooting Establishing Demand for Public Launching Facilities
	WR4:	*New Opportunities and Facilities for Watersports*		
	WR5:	*Sheltered Waters and Watersports Uses*		
	WR6:	*Public Safety*	**WR6.1:**	Monitoring Numbers of Leisure Craft
	WR7:	*Resolving Conflicts Between Water-based Recreation and Other Uses*		

Other Relevant Material

The following material produced by the Study Team is also relevant to this chapter:

TOPIC REPORTS

1 Initial Consultation with Voluntary Organisations and the Private Sector

4 Informal Recreation Opportunities

8 Water-based Recreation

OVERVIEW REPORTS

Mersey Estuary Management Plan : First Year Report

Mersey Estuary Management Plan : Report on the Consultation Exercise

MAP 3 RECREATION

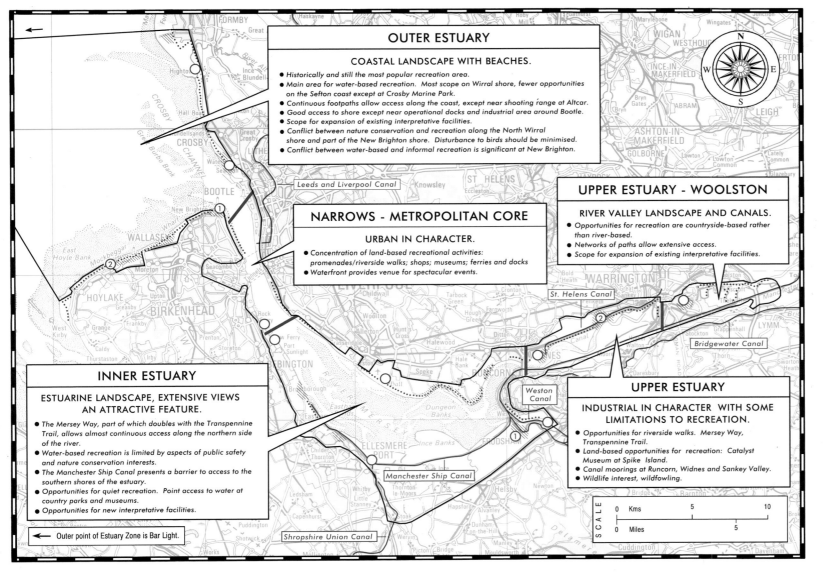

OUTER ESTUARY

COASTAL LANDSCAPE WITH BEACHES.

- *Historically and still the most popular recreation area.*
- *Main area for water-based recreation. Most scope on Wirral shore, fewer opportunities on the Sefton coast except at Crosby Marine Park.*
- *Continuous footpaths allow access along the coast, except near shooting range at Altcar.*
- *Good access to shore except near operational docks and industrial area around Bootle.*
- *Scope for expansion of existing interpretative facilities.*
- *Conflict between nature conservation and recreation along the North Wirral shore and part of the New Brighton shore. Disturbance to birds should be minimised.*
- *Conflict between water-based and informal recreation is significant at New Brighton.*

NARROWS - METROPOLITAN CORE

URBAN IN CHARACTER.

- *Concentration of land-based recreational activities: promenades/riverside walks; shops; museums; ferries and docks*
- *Waterfront provides venue for spectacular events.*

UPPER ESTUARY - WOOLSTON

RIVER VALLEY LANDSCAPE AND CANALS.

- *Opportunities for recreation are countryside-based rather than river-based.*
- *Networks of paths allow extensive access.*
- *Scope for expansion of existing interpretative facilities.*

INNER ESTUARY

ESTUARINE LANDSCAPE, EXTENSIVE VIEWS AN ATTRACTIVE FEATURE.

- *The Mersey Way, part of which doubles with the Transpennine Trail, allows almost continuous access along the northern side of the river.*
- *Water-based recreation is limited by aspects of public safety and nature conservation interests.*
- *The Manchester Ship Canal presents a barrier to access to the southern shores of the estuary.*
- *Opportunities for quiet recreation. Point access to water at country parks and museums.*
- *Opportunities for new interpretative facilities.*

← Outer point of Estuary Zone is Bar Light.

UPPER ESTUARY

INDUSTRIAL IN CHARACTER WITH SOME LIMITATIONS TO RECREATION.

- *Opportunities for riverside walks. Mersey Way, Transpennine Trail.*
- *Land-based opportunities for recreation: Catalyst Museum at Spike Island.*
- *Canal moorings at Runcorn, Widnes and Sankey Valley.*
- *Wildlife interest, wildfowling.*

Leeds and Liverpool Canal

St. Helens Canal

Bridgewater Canal

Weston Canal

Manchester Ship Canal

Shropshire Union Canal

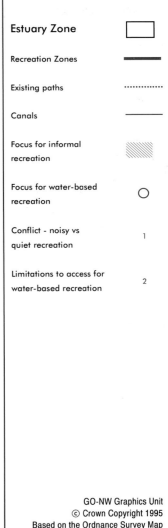

Estuary Zone	
Recreation Zones	
Existing paths	
Canals	
Focus for informal recreation	
Focus for water-based recreation	
Conflict - noisy vs quiet recreation	1
Limitations to access for water-based recreation	2

GO-NW Graphics Unit
© Crown Copyright 1995
Based on the Ordnance Survey Map

10 : Understanding and Monitoring

The implementation of management plans requires the provision of appropriate management information. Without knowledge and understanding of what is happening, and how and why it is happening, management efforts may easily become frustrated or produce unforeseen adverse after-effects. Without a good information base it is possible to miss the emergence of new issues before they become urgent and to pursue other matters long after their nature has been amended. A management plan is only good if it can be modified on the basis of new information and management objectives.

Estuary management plans, particularly those for an estuary so complex in its variety of uses and public agency interests as the Mersey, and their effective implementation, have distinctive information requirements related to these characteristics. One requirement is for highly specialist and technical information; another is for the integrated analysis of different types of specialist information; a third need is for some additional or new sorts of information or basic enquiry. Finally, there are needs to make information more widely available, to make the technical more generally understandable, and, to relate the Management Plan's policies to the changing physical, social and economic environment of the Estuary as a whole.

Four management information components have emerged from the Mersey Estuary Management Plan Study and are discussed more specifically below:

(i)   a bringing together of existing information into a database which will be regularly updated, will provide the source for more integrated studies between different specialist topics, and, provide a marker for the establishment of an Estuary research programme;

(ii) using the database as a means of regularly monitoring the state of the Estuary, its use and management so that the Management Plan's strategic objectives and policies can be evaluated and, from time to time, reviewed;

(iii) establishing an annual 'State of the Estuary' reporting system for discussion by the key policy sectors and decision agents concerned with estuary management - the local authorities, central government and public agencies, the business sector, and community and voluntary interests;

(iv) provision for environmental education and interpretation which will encourage a wider and deeper appreciation of the Estuary Zone's natural and physical resources and will contribute to a better understanding of estuary management issues, policies and processes.

OBJECTIVE *Steps should be taken to provide adequate management information to develop understanding and awareness of the natural dynamics of the Estuary and the interrelation of social and economic activities - including the objectives and policies of the management planning process - with these natural factors.*

POLICIES

Preparation and Upkeep of a Comprehensive Database

The establishment of a comprehensive multi-disciplinary base of understanding of the Estuary is regarded as an essential requirement for the practical implementation of the Management Plan. This will involve a number of connected features including:

- An inventory of currently available information;

- The specification and prioritisation of gaps in existing knowledge about the Estuary and our understanding of its processes;

- The formulation and funding of a coordinated Estuary research programme;

- A regular updating of the inventory, which would enable changes over time to be assessed and contribute to policy formulation and project assessment;

- The routine monitoring of the implementation of the Management Plan itself and the evaluation of the effectiveness of its policy objective and measures;

- The development of an annual 'State of the Estuary' reporting system.

The availability of information on the Estuary as a whole is patchy, but this does not mean that it is inconsiderable in terms of more specialist and partial information. The first task in establishing a comprehensive database is to make contact with existing data holders and establish the nature and scope of their holdings and take steps to draw this existing information together. Action should then be considered to fill any identified gaps; here there may be support for specific research proposals. A third task will work in cooperation with existing data collectors to maintain appropriate regular updates of the base, especially where series of data can be usefully maintained or need to be established.

In that sense there will be a close connection between the updating of the database and routine monitoring activities. The database is intended for use as a source of management information and there will be a need for attention to be given to the provision of appropriate reports and other publications.

UM1 *A comprehensive Mersey Estuary database, which will be updated regularly, should be established under the auspices of the Management Plan implementing organisation. The database should be accessible to all those interested in the Estuary and its working.*

The database could, for example, include information on the following aspects of the Estuary, its dynamics, use and policies: air quality; chemistry (including physical chemistry); development proposals; ecology (including biology and habitats); economics; engineering; geology and geomorphology; hydrology; navigation; public policies; recreation; sedimentology; tourism; water quality. The use of Geographic Information Systems (GIS) is now growing rapidly and its use in relation to Estuary data and management information would be advantageous.

Research priorities should be established for the Estuary and its management and a scientific research programme formulated. The programme should include social, economic and policy matters as well as those relating to the natural dynamics and engineering of the Estuary Zone. An early priority should be the review of existing research and monitoring programmes recommended in policies ED1 and PC1.

Attention would have to be given to the funding and location of the database, and, arrangements for access and use. As well as drawing on the facilities and resources of the lead agencies and other key bodies, the

The National Rivers Authority's coastal survey vessel 'Coastal Guardian' is used for monitoring water quality in the Mersey Estuary and the coastal waters of the Irish Sea

implementing organisation should draw on the full range of sources of help which may be available from the voluntary and community sector for local and specialist aspects of database management.

Monitoring and Review of the Management Plan

The use of monitoring systems in management and planning has grown rapidly in recent years with the appreciation that the process of implementing strategies, objectives and polices is as important as the preparation of the plan statement.

Monitoring implies a routine of keeping up to date with changes in a baseline record of key features, issues and indicators. It thus acts as a feedback mechanism for decision makers and others involved in a management process. The type of monitoring programme required has to be devised to fit individual circumstances and because it is costly, requires setting priorities so that desired quality standards are achieved.

Monitoring is usefully linked to the more periodic evaluatory procedure of reviewing accumulated information against approved strategy, objectives and policies. This provides an opportunity to evaluate the effects of management actions over a medium-term period (typically 3-5 years), to examine movements in trends, and, to assess the significance of new issues and changes in the physical environment, the economy, social conditions and policy preferences. This will enable discussion to proceed on confirming or revising Management Plan objectives, policies and measures and the formulation of new elements.

Two examples of useful mechanisms for monitoring and review are information systems and assessment procedures. The initial step in establishing a systematic framework for the collection and evaluation of data on the quality of the Estuary environment has been discussed in Policy UM1 above. Additional suggested steps are indicated below.

The second useful mechanism is the comprehensive and systematic assessment of the consequences of an existing or proposed action on the environment (natural, built, economic, social). Such assessments are useful ways of incorporating non-financially costed elements, which otherwise often get over-looked in decision making, into policy debate. They also provide an agenda for discussion and are increasingly being used as a standard procedure for significant proposed actions (e.g. development projects such as the proposed airport expansion at Speke) and are discussed elsewhere in the Management Plan. Although usually carried out on an ad hoc basis, they nonetheless form a useful input to the overall monitoring and review process.

UM2 *The implementing organisation should be responsible for the routine monitoring of the implementation of the Management Plan and a periodic review of its objectives and their achievement.*

Management Measures

UM2.1 *The implementing organisation should establish a priority set of Estuary characteristics and indicators for monitoring, including the natural dynamics of the Estuary and the inter-relationships with the social and economic activities of its users and the proposals of the Management Plan. An early priority should be the review of existing research and monitoring programmes recommended in policies ED1 and PC1.*

UM2.2 *The implementing organisation should prepare periodic general and occasional specialist reports on its work, and, in consultation with other components of the implementing organisation might commission occasional special studies to fill major gaps or deal with newly significant questions.*

UM2.3 *The implementing organisation shall organise a periodic review of the objectives, policies and effectiveness of the Management Plan, and, shall recommend such changes as appear desirable to the implementing organisation's executive body for consideration.*

An Annual 'State of the Estuary' Reporting System

Many organisations and agencies produce annual reports on their immediate past activities, current issues, and, their future prospects. This is a useful way of bringing information together and focusing attention for the record, for discussion and for evaluating the effectiveness of stated policies and measures. The Merseyside Development Corporation, for instance, produces an annual report together with its accounts. This report focusses on the activities of one organisation. Another type of annual report which draws information together on a range of different events and activities is that on The State of the Navigation of the River Mersey produced by the Acting Conservator of the River Mersey.

While both types of report are useful, the Annual Report on the State of the Navigation suggests the scope for a more general periodic report. This could usefully draw on information in reports prepared by other agencies during the year as well as complementary work derived from the Estuary database and related research technical, monitoring and policy studies. This would complement existing reporting systems, but also serve as a means of raising issues and bringing discussion questions to the different components of the implementing organisation and the interested general public. An annual 'State of the Estuary' report would provide the basis for an Estuary annual general meeting or conference which would bring together representatives of all the interests in the Estuary and its management.

UM3 *The implementing organisation should produce an annual 'State of the Estuary' report, drawing attention to developments and changes during the year, any matters of current concern, and, issues or proposals of future interest to the management of the Estuary.*

Management Measure
UM3.1 *An annual 'State of the Estuary' report shall be presented to the Mersey Estuary Forum for discussion. The report shall be designed for a broad-based readership including interested members of the general public.*

Environmental Education and Interpretation

The Mersey Estuary is highly significant in the history and economic development of the region. It is therefore essential that easily understood information should be made available to the public, and to students in particular, in order to encourage an understanding of its complexity and importance.

This public awareness could be promoted through the development of interpretive facilities at strategic locations, small interpretation centres and particularly through specially-designed displays at museums - such as the Liverpool and Maritime Museums and the Ellesmere Port Boat Museum. As well as 'aquarium events', such displays, given resources, could possibly employ new techniques, such as fibre optics and 'cave technology'. Aspects of the Estuary to be interpreted could include natural history and wildlife, water quality, maritime, wartime and canal history, economic development as well as local art and culture with an estuarial theme.

A programme of environmental education would promote awareness of the progress in improvement of water quality, not only among the residents of the study area and the North West, but also among potential tourists and investors which would be to the benefit of the whole estuary area.

Many features of the Management Plan will provide highly relevant issues for students from a wide range of educational establishments to study - from primary school projects to advance research in Higher Education - as well as in adult and continuing education. INSET courses for teachers linked to the national curriculum could be encouraged. The North West Environmental Education Forum could be utilised as a linkage with relevant educational organisations and appropriate educational advisors/inspectors.

Environmental education and interpretation are an essential and integral element of the management process in ensuring public understanding and commitment.

UM4 *Liaison with the Mersey Basin Trust's Education Team should be an important element in the environmental education programme envisaged and should be linked with the teaching resources generated through the Trust's educational remit.*

Figure 7: Implementation: Summary of Policy Areas, Policies and Management Measures.

Strategic Policy Area	Strategic Policy	Management Measure
10 UM: Understanding and Monitoring	**UM1:** *Preparation and Upkeep of a Comprehensive Database*	
	UM2: *Monitoring and Review of the Management Plan*	**UM2.1:** Setting Priorities for Monitoring **UM2.2:** Reporting the Results of General and Specialist Studies **UM2.3:** Periodic Review of the Plan
	UM3: *An Annual 'State of the Estuary' Reporting System*	**UM3.1:** Disseminating the State of the Estuary Report
	UM4: *Environmental Education and Interpretation*	

Organisational Arrangements for Implementation

This section sets out the Estuary Project Group's proposals for future organisational arrangements for the implementation of the Plan. It differs from the remainder of the document in two important ways. First, it has not been subject to the same consultation process as the Plan itself; the consultation draft of the Plan did not contain a section on organisational arrangements for implementation. Secondly, the proposals have originated from the Estuary Project Group; development of an implementation structure was not part of the University Study Team's commission.

As part of its commission, however, the Study Team examined experience elsewhere in estuary planning and in related areas of environmental management. With the benefit of this background material, the Estuary Project Group has concluded that the best way forward is to maintain and strengthen existing organisational arrangements under the umbrella of the Mersey Basin Campaign. The Campaign has, over the past ten years, built up an enviable reputation for promoting and implementing environmental improvements and has fostered a unique network of productive relationships in the public, private and voluntary sectors.

Implementation proposals were discussed by delegates at the 1995 Estuary Conference at which the final draft Plan was presented. Further proposals, revised as a result of comments received, have been discussed through the Estuary Project Group with local authority and other partners.

The Mersey Basin Campaign

It is considered important to reinforce the linkage between the Estuary Plan and the Mersey Basin Campaign. The Campaign covers the entire catchment of the River Mersey and its Estuary and thus has the potential to reinforce action taken on the Estuary by influencing activities within the whole catchment.

The Mersey Basin Campaign is a partnership of public, private and voluntary sector bodies and interests which aims, over a twenty five year period, to clean up the Mersey River system. The Campaign was launched in 1985 by the Department of the Environment, with the following objectives:

- To improve river quality to at least Grade 2 (fair) standard by the year 2010 so that all the rivers and streams are clean enough to support fish;
- To stimulate attractive waterside developments, for business, recreation, housing, tourism and heritage;
- To encourage people living and working in the Mersey Basin to value and cherish their watercourses and waterfront environments.

The Campaign is structured around three central components, which are represented on a monthly board under the Campaign Chairman. These are:

- The Mersey Basin Campaign Unit of the Department of the Environment, which acts as overall coordinator and administrator for the Campaign's resources and projects;
- The Mersey Basin Trust, formed as a charity in 1991 from the Campaign's Voluntary Sector Network. The Trust co-ordinates the activities of voluntary groups in undertaking environmental improvement projects and promoting education and awareness;

- The Mersey Basin Business Foundation, which comprises businesses actively supporting the Campaign through injecting business thinking as well as giving financial support for projects.

In addition to these central components to the Campaign there are also five area-based Project Groups, which represent the interests of the Campaign on the ground. The Groups are split geographically into the Upper, Central and Southern Catchments, the Estuary and the Leeds-Liverpool Canal Corridor. The Project Groups, which are led by local authorities and contain members from all sectors and interests of the partnership, act as coordinators for more locally-based projects such as River Valley Initiatives. The Mersey Estuary Management Plan falls under the wing of the Estuary Project Group.

Proposed Organisational Arrangements

The proposals for implementation build on and develop the arrangements that have operated during the preparation of the Management Plan. The proposed organisational structure, shown in Figure 8, is deliberately simple, but is capable of evolving in future, if and when further adaptation is felt to be necessary.

A central feature of the arrangements is the Mersey Estuary Forum, a development of the successful Annual Conferences which have been held throughout the plan preparation period. The Forum, which is intended to meet annually, offers the opportunity for all interested parties to be involved. This is the arena for elected members of local authorities and for senior members of other organisations. However, unlike the previous Annual Conferences which have focussed entirely on the Management Plan, the Forum will aim to cover all the work of the Estuary Project Group, including the River Valley Initiatives (such as Alt 2000). Parallel sessions for each initiative could cover detailed business, with plenary sessions providing for an exchange of information and experiences across the whole range of Estuary and river-related activity.

The Mersey Estuary Project Group would continue in its present form, bringing together officers from the forty or so organisations that have responsibilities for, and detailed interests in, the Estuary. It would include representatives from the voluntary and private sectors, as well as statutory bodies. The Project Group would meet quarterly, and like the Forum, would deal with matters related to River Valley Initiatives as well as the Estuary Management Plan.

The proposed Steering Group for the Mersey Estuary Management Plan would parallel those already in existence for the River Valley Initiatives. The Technical Steering Group which guided the preparation of the Management Plan was generally regarded as a valuable body, in large part because it was a compact, close-knit group of individuals committed to the concept of the Plan.

The new Steering Group to progress implementation of the Plan would be similarly compact, while at the same time embracing the full range of sectoral interests. The proposal is for a Steering Group comprising officers from the public, private and voluntary sectors, but it does not provide for representation from each of the individual organisations funding implementation of the Plan. Members of the Steering Group would be drawn from the membership of the Estuary Project Group.

More specifically, it is proposed that the Steering Group should comprise :
- one officer from local authorities in the Outer Estuary;
- two officers from local authorities in the Inner Estuary;
- one officer from local authorities in the Upper Estuary;
- two officers from statutory agencies;

- two officers from the private sector (including one from the Mersey Basin Campaign Business Foundation);

- two officers from the voluntary sector (including one from the Mersey Basin Trust);

a total of ten members. They would in all cases represent the interests of their sector rather than just their own organisations. The Steering Group would have the power to co-opt up to two additional people, should a specific project require specialist input not already available in the Group.

The Estuary Coordinator would function as a professional/technical advisor reporting to the Steering Group and acting as a contact point, facilitator and source of information about the Estuary and its management. The person appointed would be unlikely to have any specific project implementation role, except in the sense of managing and coordinating studies agreed by the Steering Group, or managing the Plan review process itself. The key role of the Coordinator would be one of networking with the many public, private and voluntary sector agencies to ensure that 'things happen' and that such actions are consistent with the aims and objectives of the Plan.

The main duties of the Coordinator would be to:

- Promote the aims and objectives of the Plan to all relevant agencies within the Estuary Project Group area, facilitating the exchange of information and good practice, developing and supporting a contact network between partner organisations, and adopting an advocacy role to extend the influence of the Plan in relation to developments within the Estuary Zone;

- Coordinate and monitor an Estuary-based action programme, identifying and agreeing key projects for inclusion in this programme, monitoring progress and proposing specific projects for business sponsorship;

- Provide advice, or identify sources of advice, on Estuary-related matters, specifying issues of common interest requiring study, and acting as a central reference point for information on the Estuary;

- Maintain an overview of experience elsewhere and of developments nationally, including current practice, relevant legislative frameworks and funding sources for the action programme.

Consultancy support is regarded as essential during the implementation phase, particularly in relation to the monitoring and review functions that have been identified. Consultants would work closely with the Estuary Coordinator and with the Steering Group, focussing on four main tasks:

- Preparation and upkeep of a comprehensive database - the University Study Team has already assembled a very substantial amount of Estuary-related information which will need to be updated and supplemented if it is to maintain its usefulness and relevance to the Plan;

- Monitoring and review of the Management Plan, recognising that changes will need to be made as new issues emerge and existing ones recede and as progress is made towards achieving the objectives and targets of the Plan;

- Establishment and operation of an annual 'State of the Estuary' reporting system, synthesising information about the Estuary and its management, for the benefit of a wider public audience, as well as the Estuary Forum and the Project Group;

- Making a contribution to the promotion of a programme of environmental education and interpretation, advising on its contents and providing assistance in the preparation of suitable teaching and exhibition material.

Such tasks are probably best undertaken by one team of consultants engaged on the basis of a retention arrangement. There are likely to be other tasks, however, that might best by carried out by limited-life task groups, including officers from partner organisations.

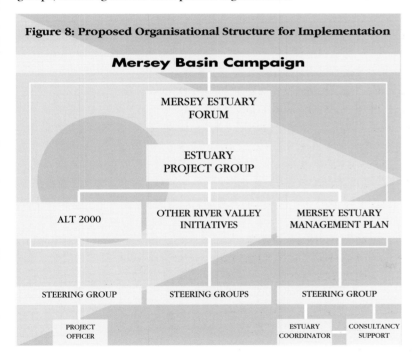

Figure 8: Proposed Organisational Structure for Implementation

Other Relevant Material

The following material produced by the Study Team is also relevant to this chapter:

TOPIC REPORTS
 5   Initial Consultation with Statutory Agencies
13   The Implementation of an Estuary Management Plan :
      Organisational Structures and Institutional Arrangements
14   Monitoring the Management Plan
15   Estuary Projects : An Outline Programme

An Index of Management Measures

1 Monitoring, Review and Information Exchange

ED1.1 Review of Monitoring Programmes in Relation to Physical Processes

PC1.1 Promoting and Disseminating Best Practice in Pollution Control

PC1.4 Agreeing Programme of Monitoring in Relation to Pollution Control

PC3.1 Survey and Assessment of Options for Contaminated Land

BD1.1 Baseline Survey of Key Sites and Habitats

BD2.1 Baseline Survey of Key Species

LU1.2 Reviewing Land Allocations for Industry

LU2.1 Assessment of Townscape

LU4.1 Assessment of Landscape

NV4.1 Undertaking a Technical Study of Port Sites

NV6.1 Assessing Dredging Obligations

NV7.2 Providing for Environmental Assessment in Areas Covered by Permitted Development Rights

RG3.1 Undertaking a Survey of Sites and Reviewing Options for Remedial Action

WR1.1 Monitoring Access for Recreation Sailing

UM2.1 Setting Priorities for Monitoring

UM2.2 Reporting the Results of General and Specialist Studies

UM2.3 Periodic Review of the Plan

UM3.1 Disseminating the State of the Estuary Report

2 Development of Programmes of Action and Setting of Targets

PC1.2 Agreeing Pollution Reduction Targets

PC3.2 Programme of Care and Remedial Treatment for Contaminated Land

BD1.2 Safeguarding Measures and Enhancement Targets for Key Sites and Habitats

BD2.2 Safeguarding Measures and Enhancement Targets for Key Species

LU1.1 Coordinating Plans within Corridors

RG3.2 Establishing Reclamation Programmes

IR4.2 Coordinating MEMP and Mersey Forest Plan in Relation to Public Access

3 Development of Guidelines and Policies

BD1.3 Plans for Key Sites, Habitats and Features

BD2.3 Plans for Key Species

LU1.1 Coordinating Plans within Corridors

LU2.2 Developing Townscape Guidelines

LU4.2 Developing Landscape Guidelines

NV7.1 Establishing an Environmental Code of Practice

4 Identification, Reservation and Promotion of Sites

PC1.3   Meeting Pollution Reduction Targets

LU1.2   Reviewing Land Allocations for Industry

NV3.1   Reserving Sites for Future Port Use

NV4.1   Defining Port-related Employment Areas

NV4.3   Encouraging Variety in Sites, Users and Programming Priorities

NV4.5   Providing for the Reclamation and Reuse of Port Sites

NV6.1   Assessing Dredging Obligations

TR1.1   Identifying Tourist Activities, Facilities and Sites
to Promote Economic Regeneration

TR2.1   Identifying and Promoting Sites for Tourist Accommodation

IR4.1   Appropriate Design and Designation of Routes

WR1.2   Relocating Facilities for Watersports and Recreation

5 Maintenance and Enhancement of Facilities and Amenities

IR2.1   Maintaining Access to the Coast

IR4.1   Appropriate Design and Designation of Routes

IR5.1   Restrictions to Public Access

IR7.1   Extending Coverage of Ranger Services

WR1.3   Maintaining and Enhancing Access and Facilities
for Water-based Recreation

6 Reinforcement of Policies and Actions of Other Agencies

PC1.3   Meeting Pollution Reduction Targets

NV4.4   Supporting the Port of Liverpool's European Gateway Role

NV5.1   Continuing and Developing Training and
Enterprise Support Schemes

RG1.1   Supporting Availability of a Portfolio of Regeneration Proposals

RG1.2   Supporting Environmental Upgrading of
Existing Developed Areas

RG2.1   Continuing and Developing Training and
Enterprise Support Schemes in Existing Developed Areas

RG5.1   Promoting Industrial, Service and Commercial Contributions
to Oil and Gas Production

RG5.2   Supporting Wider Activities to Examine Implications
of Oil and Gas Licensing, Exploration and Production

TR1.1   Identifying Tourist Activities, Facilities and Sites
to Promote Economic Regeneration

TR6.1   Encouraging Relevant Agencies to Promote
Tourist Attractions of the Estuary

IR4.2   Coordinating MEMP and Mersey Forest Plan
in Relation to Public Access

APPENDIX 2

Assessing Development Projects:
Three Hypothetical Examples

This section is intended to illustrate how the Management Plan policies may be applied in assessing development projects. Three hypothetical examples have been chosen : a road bridge crossing; the redevelopment of Princes Dock; and a new Estuary interpretation and recreation centre, with an associated ferry landing stage. The examples have been chosen to represent different scales of development and different kinds of impact. In this way, a range of policies may be tested for their relevance and utility.

A common approach has been followed in describing the examples. After introducing the proposal, a broad indication of the likely impacts is given. Then the Management Plan policies are applied, and finally implications for the decision-making process are outlined.

Case Study 1:
Road Bridge Crossing of the Mersey

Introduction

The proposal considers the development of a road bridge crossing the Mersey, an option initially explored by Merseyside Area Land Use Transportation Study (MALTS) in the late 1960's. The favoured route for the bridge put forward by MALTS was Magazine Road, Bromborough Port to Jericho Lane, Otterspool.

Likely Effects

Listed below are the likely effects that would need to be considered in relation to the proposal:-

ESTUARY DYNAMICS

- Siting of bridge's pier supports could affect the physical regime of the Estuary
- Extra flood defence works could be required.

WATER QUALITY & POLLUTION CONTROL

- Precautions would be needed to avoid pollution of the Estuary during the construction period.
- Siting of pier supports could disturb sediments and release heavy metals.
- Adverse effects could occur as a result of polluted surface water run-off from the bridge.

BIODIVERSITY

- Changes in water quality could affect the ecology of the local marine environment around the bridge supports.
- There could be effects on the size and quality of estuarine habitats e.g. mudflats.
- There could be effects on the quality of the adjacent SSSI; significant effects on birdlife, particularly with regard to their flight paths.

LAND USE AND DEVELOPMENT

- There may be alterations to the Estuary vista and waterfront views, visual dominance of the bridge and disturbance to the open character of the sites.

COMMERCIAL NAVIGATION AND PORT DEVELOPMENT

- Physical regime changes could affect navigation access for the Associated British Ports site at Garston Docks and for the Manchester Ship Canal; may lead to a requirement for dredging.

URBAN REGENERATION

- There could be pressure for new economic development in the areas next to the bridge.
- The proposal could strengthen economic development and cultural activities across the Estuary.

TOURISM

- The bridge could be seen as an attractive gateway to Liverpool/Wirral where visitors could enjoy the Estuary while crossing the bridge. It could have a direct and negative impact on the Mersey Ferries business.

INFORMAL AND SHORE-BASED RECREATION

- There would be increased access adding to the network of recreation links.
- Noise disturbance during construction would affect quiet recreation close to bridge.

WATER-BASED SPORT AND RECREATION

- Siting of bridge's pier supports could hinder certain sport and recreation activities.
- The proposal would increase access to water-based sport and recreation by enabling cross river traffic.

Case Study 1: Policy Application

ESTUARY DYNAMICS

Policy **ED1**: *MEMP and Specialist Studies;* the expected effects on the Estuary's physical regime would require a specialist study, this could form part of an Environmental Assessment. The proposal would uphold Measure **ED1.1**: *Review of Monitoring Programmes.*

Policy **ED2**: *Estuary Dynamics & New Development;* Attention should be paid to the ability of the Estuary to function as naturally as possible.

WATER QUALITY AND POLLUTION CONTROL

Policy **PC1:** *Pollution Control : Best Practice Measure*
PC1.4: *Agreeing Programme of Monitoring* would support a
temporary monitoring programme of the pollution controls.

Policy **PC2:** *Pollution Control and New Development (iii)* refers to
pollution effects during construction, during this phase the main effects
on water quality would be likely to occur. Attempts should be made to
minimise expected pollution.

BIODIVERSITY

Policies **BD1:** *Site and Habitat Protection* and
BD2: *Species Protection;* apply to protect the SSSI, proposed Ramsar
Site, Sites of Local Biological Interest and Regionally Important
Geographical and Geomorphological Sites.

Policy **BD3:** *Creative Conservation;* could be used to enforce mitigating
measures in order to reduce any effects identified.

LAND USE AND DEVELOPMENT

Policy **LU1:** *Development within the Estuary Zone;* would discourage
siting the bridge across open coast and prefer siting in a developed area.

Policies **LU2:** *Built Environment Promoting Design Quality* and
LU4: *Promoting Landscape Quality* - retaining the open coast would
be relevant here. The proposal emphasises the value of
Measure **LU2.1:** *Assessment of Townscape.*

COMMERCIAL NAVIGATION AND PORT DEVELOPMENT

Policy **NV2:** *New Developments and Shipping Access;* would be
relevant here.

Policy **NV5:** *Port Related Business and Enterprise* would be in
favour of the bridge as it would help such enterprises by improving links.

URBAN REGENERATION

Policy **RG1:** *Stimulating Waterside Regeneration* would be compatible
with the bridge as it would concur with
Measures **RG1.1:** *Supporting a Portfolio of Regeneration Proposals*
the bridge being a 'flagship' scheme and **RG1.2** the bridge improving the
infrastructure of the Estuary Zone.

Policy **RG2:** *Regeneration Related Business Enterprise* would favour
the bridge as commercial concerns would benefit.

TOURISM

Policy **TR1:** *Promoting and Maintaining Diversity in Tourist
Attractions;* the bridge could become a tourist attraction.

Policy **TR4:** *Enhancing the Role of the Mersey Ferries;* the bridge
would compete with the Mersey Ferries for cross-river traffic.

Policy **TR6:** *Reinforcing Links Between Tourist Attractions;* would
uphold the proposal as it would facilitate greater access between tourist
attractions.

Policy **TR7:** *The Value of Special Events;* the opening of the bridge would be such an event.

INFORMAL AND SHORE-BASED RECREATION

Policy **IR3:** *Public Access and Waterside Developments;* the bridge would uphold this policy as it includes crossing for pedestrians, cyclists etc.

Policy **IR5:** *Public Safety;* separating pedestrians and cyclists from vehicles would be possible.

Policy **IR6:** *Encouraging the Use of Paths and Open Spaces;* the bridge would encourage use of coastal and riverside walkways in both open and built up areas.

WATER-BASED SPORTS AND RECREATION

Policy **WR1:** *Retaining Existing Facilities* would apply as pier supports could affect Estuary dynamics. The bridge would reinforce the case for Measures **WR1.1:** *Monitoring Access for Recreation Sailing* and **WR1.3:** *Maintaining and Enhancing Existing Access.*

Decision Framework

The Mersey Estuary Management Plan serves as a good basis for considering the above example, and as such could be used by the range of agencies involved in the planning application, providing a degree of common ground for their assessment. The Mersey Estuary Management Plan is also potentially beneficial in that it enables secondary effects to be considered along with several issues which are currently outside the statutory planning framework.

Case Study 2:
Princes Dock Mixed Use Development

Introduction

The development proposals consist of 500,000 sq ft of multi-storey and low-rise offices, 100 residential apartments, a hotel facility, ancillary retailing and an entertainment/restaurant facility. The waterspace is to be retained and pedestrian links with adjoining areas are to be incorporated. Mersey Docks and Harbour Company and P&O Properties have formed a Princes Dock Development Company in order to promote the redevelopment scheme. The proposed location for this development is redundant dockland north of the Pier Head.

Likely Effects

Listed below are the likely effects that should be considered in relation to the proposal:-

ESTUARY DYNAMICS

- There would be a need to ensure that adequate flood defences are in place.

WATER QUALITY & POLLUTION CONTROL

- There would be a need for precautions to avoid pollution of the Estuary during the construction period.
- There would be a need to address potential effects of surface water drainage.
- The provision of all the necessary infrastructure for water supply and drainage should prevent pollution.

BIODIVERSITY

- The waterspace could be used to create new habitats for wildlife.

LAND USE AND DEVELOPMENT

- The proposal could alter the physical character of the area, enhancing/complementing the built environment.

COMMERCIAL NAVIGATION AND PORT DEVELOPMENT

- Princes Dock would be unable to return to its traditional uses.

URBAN REGENERATION

- Redevelopment would contribute to the regeneration objectives of the Estuary Zone and provide improvements in environmental terms.
- The visual impact of the proposal would be a key consideration due to its strategic position.
- The effect of the design in the wider context of the City's historic buildings would need to be considered.
- Retaining the waterspace feature would help maintain the distinctive visual identity of the area.
- The proposal would be likely to attract new investment.
- The effects of increased traffic, parking and congestion would have to be considered.

TOURISM

- The proposal would provide accommodation for visitors and hence support existing facilities.

INFORMAL AND SHORE-BASED RECREATION

- The proposal would provide an opportunity to expand public rights of way along the waterfront.
- There would be a need to consider design and safety of a public right of way.

Case Study 2: Policy Application

ESTUARY DYNAMICS

Policy **ED2:** *Estuary Dynamics and New Development;* Attention would have to be paid to flood defence requirements.

WATER QUALITY AND POLLUTION CONTROL

Policy **PC2:** *Pollution Control and Development* would require adequate water supplies and drainage facilities and appropriate pollution prevention measures during construction.

BIODIVERSITY

Policy **BD3:** *Creative Conservation* would suggest the creation of new habitats in the waterspace and along the pedestrian links helping to establish wildlife corridors.

LAND USE AND DEVELOPMENT

Policy **LU1:** *Development within the Estuary Zone* supports the proposal as it would fall within an existing developed area.

Policies **LU2:** *Built Environment* and **LU3:** *Retaining Waterfront Heritage* would apply to this proposal which supports Measures **LU2.1:** *Assessment of Townscape* and **LU2.2:** *Developing Townscape Guidelines.* Good quality design both in terms of form, layout and buildings would be essential to ensure that new proposal complements the existing visual character of the waterfront.

Policy **LU4: *Promoting Landscape Quality*** would apply as possibilities for enhancement and informal recreation could exist.

COMMERCIAL NAVIGATION AND PORT DEVELOPMENT

Policy **NV4: *Port Related Employment Areas*** would require an examination of alternative (i.e. port related) uses for the dock. The proposal reinforces the case for Measures **NV4.1: *Defining Port-related Employment Areas,* NV4.2: *Undertaking a Technical Study of Sites*** and **NV4.3: *Encouraging Variety of Sites.***

URBAN REGENERATION

Policy **RG1: *Stimulating Waterside Regeneration*** supports the proposal as do Measures **RG1.1: *Supporting Availability of a Portfolio of Regeneration Proposals*** as the site is a 'flagship' site and **RG1.2: *Supporting Environmental Upgrading of Existing Developed Areas.***

Policy **RG2: *Regeneration and Business Related Enterprise*** supports the proposal as the proposed regeneration is in an existing developed area.

Policy **RG3: *Reclaiming and Reusing Derelict Land and Buildings*** supports the proposal as the site is presently redundant dockland. The proposal upholds Measures **RG3.1: *Undertaking a Survey of Sites and Reviewing Options for Remedial Measures*** and **RG3.2: *Establishing Reclamation Programmes.***

TOURISM

Policy **TR2: *Development of New Facilites for Tourists;*** the promotion of such facilities would be likely to attract overseas and domestic visitors. Proposals of this kind help encourage visitors to stay longer and take advantage of the tourist and recreational facilities within the Estuary Zone.

Policy **TR3: *Securing a Critical Mass of Tourist Services;*** the proposal would encourage investment and therefore support this policy.

Policy **TR6: *Reinforcing Links between Tourist Attractions;*** the proposal would assist in reinforcing links between tourist attractions by providing overnight or short stay accommodation.

Policy **TR7: *The Value of Special Events;*** the facilities would help the staging of such events and the opening of the development couldconstitute an event in its own right.

INFORMAL AND SHORE-BASED RECREATION

Policies **IR2: *Access on the Built up Coast*** and **IR3: *Public Access and Waterside Development*** would apply as the proposal would promote rights of way and provide access to the bankside.

Policy **IR4: *Extending the Network of Paths;*** the proposal would extend the existing network of paths.

Policies **IR5: *Public Safety*** and **IR6: *Encouraging the Use of Paths and Open Spaces;*** both would endorse the proposal which emphasises safety and accessibility.

Decision Framework

The policies within the Mersey Estuary Management Plan which are relevant to this case study could be integrated within the Unitary Development Plan. This would bring the policies within a statutory framework and give them additional weight.

Case Study 3:
Estuary Interpretation and Recreation

Introduction

This development proposal considers the location of an Estuary Interpretation and Recreation Centre, located within the Inner Estuary, in the grounds of Speke Hall and adjacent land, forming part of the Liverpool Northern Airfield. A new visitor centre will be located close to the waterfront. Creative habitats will be located within the grounds, including footpaths and nature trails. A new ferry landing stage will be necessary to facilitate a ferry service and nature cruiser. The centre could also be used as a base for the Ranger Service. The existing road route to the centre may need upgrading to accommodate extra traffic.

Likely Effects

Listed below are the likely effects that should be considered in relation to the proposal:-

ESTUARY DYNAMICS

- Construction of the land stage could affect the physical regime of the Estuary. Increased activity may lead to coastal erosion. Dredging may be required to improve access for ferries; this may disrupt the local ecology.

WATER QUALITY & POLLUTION CONTROL

- Dredging would disturb the river bed and effect water quality.
- Boating activity could generate pollution.

BIODIVERSITY

- Could disturb plant and animal life in the adjacent SSSI.

LAND USE AND DEVELOPMENT

- There would be some visual impact of Interpretation Centre and car park on the open shore.

COMMERCIAL NAVIGATION AND PORT DEVELOPMENT

- Ferry service would lead to increased traffic; need to consider safety aspects.
- Dredging may have an effect on existing routes through disturbance of river bed.

URBAN REGENERATION

- Proposal would bring derelict land at the Northern Airfield into use.

TOURISM

- The centre would serve as major tourist attraction, drawing people into the area.
- The proposal might increase the viability of the Mersey Ferries by providing an extra route.

INFORMAL AND SHORE-BASED RECREATION

- There would be new opportunities to gain access to the natural estuarine coast.

SPORT AND WATER-BASED RECREATION

- Access would be increased via the Ferry service and landing stage.

UNDERSTANDING AND MONITORING

- The centre would serve as a valuable educational resource.

Case Study 3: Policy Application

ESTUARY DYNAMICS

Policy **ED1:** ***MEMP & Specialist Studies*** could be used to discover expected effects of the proposal, in particular dredging. The proposal emphasise the case for Measure **ED1.1:** ***Review of Monitoring Programmes.***

Policy **ED2:** ***Estuary Dynamics and Development*** would apply to ensure that the centre is located outside any flood risk areas and attention would have to be paid to changes in the functioning of the Estuary as a result of dredging.

WATER QUALITY AND POLLUTION CONTROL

Policy **PC2:** ***Pollution Control and Development;*** attempts should be made to minimise water pollution as a result of the proposal both during and after construction.

Policy **PC3:** ***Contaminated Land*** would support the of contaminated land for the development.

BIODIVERSITY

Policies **BD1**: *Sites and Habitats Protection* and
BD2: *Species Protection* would apply to protect the SSSI and proposed
Ramsar Site, such policies are at the heart of the centres aims and the
proposal would support Measures **BD1.1** & **BD2.1**: *Baseline Surveys*
and **BD1.2** and **BD2.2** *Safeguarding Measures and Enhancement
Targets* as well as **BD1.3**: *Meeting Pollution Reduction Targets* and
BD2.3 *Plans for Species* for management and action plans for the
conservation status of sites and habitats and species.

Policy **BD3**: *Creative Conservation;* would support the development
and suggest retaining natural features and create other habitats.

LAND USE AND DEVELOPMENT

Policy **LU1**: *Development within the Estuary Zone;* the proposal
would conflict with the policy given its location.

Policy **LU2**: *Built Environment;* the Interpretation Centre would need to
be designed to harmonise with Speke Hall.

Policy **LU3**: *Retaining Waterfront Heritage;* would highlight the need
for sensitive design of new developments.

Policy **LU4**: *Promoting Landscape Quality;* would oppose the
development as it would be visually intrusive in an area of open coast. A
flexible interpretation could mean that new elements should be
sensitively designed; the visual impact of the visitor centre should be
minimised.

COMMERCIAL NAVIGATION AND PORT DEVELOPMENT

Policy **NV2**: *New Developments and Shipping Access* would be
relevant as the change in physical regime and the nature cruises may
prejudice maintenance of access by shipping using existing navigation
channels.

Policy **NV6**: *Dredging Obligations & Disposal* and
Measure **NV6.1**: *Assessing Dersging Obligations* would apply to the
proposal as ferry access could affect such obligations.

URBAN REGENERATION

Policy **RG1**: *Stimulating Waterside Regeneration* would support the
proposal as an Estuary location is required.

Measures **RG1.1** and **RG1.2** would endorse the proposal as it would be a
'flagship' scheme and involve an improvement in infrastructure.

TOURISM

The proposed development is supported by Policy
TR1: *Promoting and Maintaining Diversity in Tourst Attractions*
and Management Measure **TR1.1** for identification of activities, facilities
and sites for inclusion in the regeneration portfolio (**RG1**: *Stimulating
Waterside Regeneration*).

Policy **TR2**: *Development of New Facilities for Tourists* would
support the proposal as improved links would also help Speke Hall

Policy **TR4**: *Enhancing the Role of the Mersey Ferries*
would support the proposal as it would create custom for the ferries.

Policy **TR5:** *Development on the Open Shores;* it is questionable whether the development could be regarded as low key, as recommended under this policy.

Policy **TR6:** *Reinforcing Links Between Tourist Attractions* would support the proposal as the ferries would improve access to Speke Hall.

INFORMAL AND SHORE-BASED RECREATION

Policies **IR2:** *Access on the Built up Coast,*

IR3: *Public Access and Waterside Development,*

IR4: *Extending the Network of Paths,*

IR6: *Encouraging the Use of Paths & Open Spaces* and

IR8: *Maintaining Opportunities for Quiet Recreation* support the development since it will promote the existing Mersey Way and lead to the creation of circular footpaths. Policy **IR8** provides strong support for the proposed development.

Policy **IR7:** *Extending Interpretation Services* is entirely consistent with the proposal.

Policy **IR5:** *Public Safety* would apply to access while Measure **IR5.1:** *Restrictions to Public Access* could apply for safety reasons.

WATER-BASED RECREATION

Policy **WR4:** *New Opportunities and Facilities for Watersports* supports the development and Measure **WR4.1:** *Establishing Demand for Public Launching Facilities* would apply.

Decision Framework

The Management Plan policies are generally supportive of this proposal. The main areas of contention are likely to be associated with the new landing stage and ferry services which may conflict with navigation and nature conservation interests. The Plan has the merit in that it can be used to guide a decision on a development involving both landward and Estuary-based activities.

Preparation of the Mersey Estuary Management Plan

The purpose of this appendix is to provide details of the background to the preparation of the Mersey Estuary Management Plan. It is divided into three sections. In the first of these the initial proposal for a Plan, dating from 1989, is presented. This began life as a paper prepared for an Ecology Consultative Group connected with the Mersey Barrage project which was being actively promoted at the time. Dr. John Handley (now at the University of Manchester, but then Director of St Helens and Knowsley Groundwork Trust) argued the case for a Management Plan, setting out a number of ideas for its scope and organisation.

These ideas were subsequently taken up by the Mersey Basin Campaign's Estuary Project Group. In 1991 the Group invited the Department of Civic Design at the University of Liverpool to take on a three-year commission to prepare a Management Plan for the Estuary. The second section of the appendix consists of the brief prepared by Liverpool City Council on behalf of the commissioning partners. It lays out the main tasks to be undertaken and provides guidelines on how the work should be organised.

The final section describes the University Study Team's approach to the commission. It outlines the role played by the University in formulating the Plan, and discusses the benefits to academic staff and to graduate students of participating in a major project of this kind.

1 The Original Proposal for a Management Plan

An abridged version of a paper prepared originally by Dr.J.F.Handley for the Mersey Barrage Ecology Consultative Group, May 1989, entitled *"Towards a Management Plan for the Mersey Estuary"*

The Need for a Management Plan

The Mersey Estuary is a natural resource of international importance for shipping and nature conservation. The Estuary has a large tidal range and tidal volume and it is the scouring effect which this creates that helps to maintain the shipping lanes of the deep water port.

The Estuary lies at the mouth of a heavily populated and industrialised catchment. It remains one of the most polluted estuaries in Europe by domestic and industrial discharges. Poor quality water destroyed the fishing industry and this, together with the tidal regime, inhibits the use of the water surface for recreation and causes fouling of foreshores. Recent investment, accelerated by the Mersey Basin Campaign, will improve water quality to a point where, by the end of this century, commercial fishing may have returned to the Estuary and recreational use will have intensified.

The potential for tourism and recreation has been shown by the work of Merseyside Development Corporation, with the Albert Dock now one of the most popular tourist destinations in Britain.

It follows that, as the water quality of the Estuary improves and the Merseyside economy redevelops, there will be increasing competition between uses and users of the water frontage and the water surface. This would be accelerated by a Mersey Barrage, should it materialise, but these issues will intensify in any event.

A Management Plan is needed to provide a framework within which development proposals can be assessed and through which activity on and around the Estuary could be coordinated. It would safeguard existing interests whilst realising the full potential of the Estuary as a natural resource. It is unlikely that this will be achieved in the absence of a Management Plan because of the many interests and agencies involved in the Estuary, each with its own concerns and responsibilities, which are in no way integrated at present.

The Management Plan - Aims and Objectives

It is proposed that the aim of the Management Plan should be:

"to provide a framework for the management of the Mersey Estuary within which existing interests can be safeguarded whilst realising the full potential of the Estuary as a natural resource."

The specification of a series of clear and achievable objectives, which could be agreed by all parties, would be an important part of the Management Plan process.

The Scope of the Management Plan

(a) DEFINITION OF THE MERSEY ESTUARY

The spatial definition of the Mersey Estuary for the purpose of the Management Plan is not straightforward. The Working Group which helped to draft this paper was inclined to take a pragmatic view which would define the Mersey Estuary as extending from Howley Weir, Warrington (the upper tidal limit) to the mouth of Narrows (between Fort Perch Rock and the Seaforth Radar Station) at the seaward end. The lateral extent of the area covered by the Mersey Estuary Management Plan would extend beyond extreme high water of spring tides. It would take in adjoining land within an Estuary Zone where land use is considerably

influenced by, or impacts on, the Estuary. In some sections, such as Hale Point, this zone may extend well inland whereas in the 'Narrows' it would equate with the maritime zone.

Other interpretations are possible. For example, for the purposes of the Mersey Basin Campaign the Estuary Zone is interpreted as taking in the North Wirral foreshore and the Sefton coastline up to, and including, the River Alt. Similarly, a definition of the Estuary which reflects the jurisdiction of the maritime authorities would extend well into Liverpool Bay.

At the upper end of the Estuary the Howley Weir boundary is equally arbitrary. In a policy for tourism and recreation developed by Cheshire County Council, in association with the District Councils, the project area extended up the Mersey Valley taking in Woolston Eyes and terminated at the administrative boundary with Greater Manchester.

(b) REGULATORY AGENCIES, MANAGING AGENCIES AND INTEREST GROUPS

One of the major obstacles to resource management of the coastal zone in Britain is that, understandably, the focus of legislation and interest is landward, rather than seaward. There is a general need for a more coherent government policy for the coastline and this is nowhere more apparent than in the estuarine environment. Development pressures here tend to be at their most intense, but the intrinsic worth of the Estuary as a natural environment often goes unrecognised. Consequently, whilst many of Britain's estuaries are scheduled as Sites of Special Scientific Interest by the Nature Conservancy Council (now English Nature), almost none are designated as Heritage Coast by the Countryside Commission.

Where the existence of an estuary is recognised officially it is often because it forms a convenient physical boundary to an administrative area. The Mersey Estuary has no less than ten Local Authorities and Development Corporations within its boundaries, each with their own planning responsibilities:

- Cheshire County Council
- Ellesmere Port and Neston Borough Council
- Halton Borough Council
- Liverpool City Council
- Merseyside Development Corporation
- Sefton Metropolitan Borough Council
- Vale Royal Borough Council
- Warrington and Runcorn Development Corporation[1]
- Warrington Borough Council
- Wirral Metropolitan Borough Council

The Local Authorities and Development Corporations are just some of the many agencies with regulatory powers which affect the Mersey Estuary. In addition to the regulatory agencies, there is a smaller group of managing agencies which are involved directly in maintaining or modifying the Estuary and its immediate surroundings. These agencies would have an especially important role to play in implementing the proposals of the Management Plan and they include:

- Local Authorities
- Merseyside Development Corporation
- Mersey Valley Partnership
- Competent Harbour Authorities
- North West Water
- Cheshire Trust for Nature Conservation
- Lancashire Trust for Nature Conservation
- Merseyside Naturalists' Association

The preparation of the Management Plan would involve the regulatory and managing agencies in particular, but would also need to take account of the views of owners and occupiers, users of the Estuary and special interest groups. This would require extensive and lengthy consultation. Representative bodies such as the Confederation of British Industry, the Mersey Estuary Conservation Group and the Estuary Project Group of the Mersey Basin Campaign, would have a key role to play in facilitating this.

The Management Plan Framework

The Management Plan will need to win support from the agencies, interest groups and owners and occupiers if it is to be effective. At its most basic the Management Plan would be no more than a compilation of existing policies and programmes drawn from amongst the regulating and managing bodies. However, this in itself would be a major step forward and it would at least point out the need for harmonisation of policies and coordination of action.

At its most developed the Management Plan would be prescriptive with an agreed work programme and annual work plan. The implementation of this work plan would be the responsibility of the managing agencies. This would be especially appropriate for a work programme related to

---

[1] Ceased to operate in September 1989

countryside management and useful models already exist in the Sefton Coast Management Scheme and the Mersey Valley Project.

The Management Plan would meet early opposition if it sought, from the outset, to assume executive powers from the regulatory and managing agencies. The Management Plan might well be similar in form to the National Park Plans which the National Park Authorities in England and Wales are required to produce. This is not a part of the development plan process, nor does it have any direct bearing upon development control procedures within the National Park. Instead, it is intended to be a management plan in which the authority sets out its National Park objectives for the whole area within the Park boundary and seeks to obtain the agreement of all concerned, including statutory and voluntary bodies owning land within the National Park, to implementation of those objectives over a five-year programme.

The Mersey Estuary Management Plan would equally seek to find common ground between the many organisations involved in providing an agreed policy framework for the Mersey Estuary. This might take the form of a Mersey Estuary Convention which, whilst it would not be legally binding because it has no basis in planning law, could be signed by all the prime movers within the Estuary. The Management Plan would be reviewed at regular intervals, perhaps every five years, at a Mersey Estuary Conference.

The preparation of a Management Plan would involve drawing together a large and complex body of information about the Estuary and its environs. This would need to be done by a small planning team housed within one of the regulatory agencies and supported by a consortium of lead agencies in cash or in kind. Besides formulating the policy framework, the planning team would identify key indicators of environmental quality which should form the basis for long term monitoring - so providing a yardstick against which the success of the Management Plan could be judged.

If it was found that progress could not be achieved through this mechanism and that the Estuary was deteriorating in quality, it would be necessary to consider an alternative management structure. Suitable models are not easy to find, but one possible approach is the Broads Authority set up to combat the falling quality of water and the habitat of the Norfolk Broads. Such a body would assume both regulatory powers and executive responsibilities making it both a regulatory and managing agency in its own right.

2 Consultant's Brief

Brief prepared by Liverpool City Council and issued to the University of Liverpool in a commissioning letter - February 1992.

Objectives

The aim of the Mersey Estuary Management Plan is to provide an advisory framework for the future management of the Mersey Estuary within which existing interests can be safeguarded and development proposals evaluated, to enable the realisation of the fullest possible potential of the Estuary as a natural resource.

Specifically, the commissioning partners intend that the Management Plan will:

- Focus attention on the Estuary as one of the Mersey region's most important environmental assets and convey a positive image of the area as a unique conurbation with an enormous water resource (with recreational and tourist potential) at its core;

- Provide the basis for an agreed and coordinated programme of environmental action and creative conservation to be implemented by the commissioning partners and others;

- Set out proposals for the management of river-based recreation and for the protection of ecological assets;

- Establish part of the technical basis to enable the Local Authorities and others to respond to major development initiatives on the Estuary;

- Enable the commissioning partners to speak with an informed and authoritative voice on matters affecting the Estuary.

Commissioning Partners

The commissioning partners for the Management Plan are the following agencies:[2]

- Department of the Environment (Mersey Basin Campaign Unit)
- National Rivers Authority
- English Nature
- Liverpool City Council
- Sefton Metropolitan Borough Council
- Wirral Metropolitan Borough Council
- Cheshire County Council
- Halton Borough Council
- Warrrington Borough Council

The authorities have come together at an officer level through the Estuary Project Group of the Mersey Basin Campaign which will act as the primary recipient for the study. On a day-to-day basis, however, Liverpool City Council, which chairs the Estuary Project Group, will act as the lead authority and client for the purposes of budgetary control, and will commission the consultants in accordance with the Council's usual procedures.[3]

---

[2] Ellesmere Port and Neston Borough Council and Vale Royal Borough Council were subsequently added to this list.

[3] Part way through the commission, this responsibility was taken over by Sefton Metropolitan Borough Council.

Steering Arrangements

In order to guide and oversee the direction of the consultant's work at a technical level the Estuary Project Group will create a small Technical Steering Group which will meet at regular intervals in order to help the consultants identify priority tasks and develop their programme of work. Members of the group have been identified in relation to the professional contribution which they can make to the task rather than as representatives of any particular organisation.

The Technical Steering Group will assist the consultants. It will not approve papers or policies. These should be submitted to Liverpool City Council as lead authority and client for consideration by the commissioning partners and the Estuary Project Group as a whole.

The consultant will be expected to make the necessary administrative and secretarial arrangements to service the functioning of the Technical Steering Group.

Timescale and Reporting Back to Clients

The client anticipates that the study will take 36 months to complete.

The consultants will be required, in addition to any technical or working papers produced, to submit a first interim report at 12 months, a second interim report at 24 months and a final report at 36 months.

It is expected that the first interim report will summarise initial information assembly and set out the consultant's approach to the exercise. The second interim report will serve as the basis for the draft final report.

The client must be satisfied with the direction of the interim reports before progressing to the next stage. The draft final report will be available for comment to the commissioning partners, through Liverpool City Council. Any agreed revisions will be incorporated into the final report.

In addition to submitting reports to meetings of the Estuary Project Group as described above, the consultant will be required to organise an annual conference presentation at three dates to be agreed with the clients within the commissioning period in order to serve as a vehicle for raising issues and for consultation with clients, elected members from participating authorities and other agencies, and for reporting back on progress.

Scope of Work

The consultant will be required to submit detailed work programmes for discussion with the Technical Steering Group.

It is however expected that the scope of work will include the following:

- Precise definition of the area which will be the subject of the Management Plan;
- Survey of the existing situation in relation to matters including land ownership, pollution, tidal regime, navigation channels, levels of use and nature conservation;
- Review of existing policies and statutory agencies;
- Review of international experience and mechanisms for estuary management;
- Identification of likely future developments and opportunities including role of the ports and recreational demands;

- Initial consultation with interested parties including user groups, trade and business interests and the voluntary sector;
- Identification of existing interests to be safeguarded;
- Identification of potential zones and opportunities for land and water based recreation, nature conservation and wildlife, and environmental education;
- Identification of a positive programme of proposals and remedial measures for consideration by the relevant agencies as part of their programmes;
- Advice on appropriate mechanisms for future management of the Estuary;
- Consultation with land owners, local authorities, statutory agencies, user groups, trade and business interests and the voluntary sector on the draft plan; and
- Revisions to draft plan.

3 The University Study Team's Approach

The relationship between the University Study Team and the commissioning partners was not that of conventional consultants working for a client. The Study Team played a central role in all aspects of plan preparation, including public consultation exercises and the organisation of annual Mersey Estuary Conferences, as well as the technical work associated with the drafting of the Plan. The Team sought to develop shared ownership of the Plan by involving as many interested parties as possible at all stages in the commission and benefited from its position as an independent body, in contrast to the vested sectoral and geographical interests that inevitably characterise many aspects of the Estuary.

The Study Team developed a work programme covering the three years of the commission and this is shown, in diagrammatic form, in Figure 9. In broad terms, the Team focussed on the identification of issues and the preparation of topic-based reports of survey in the first year; on the development of initial proposals for inclusion in a Draft Plan in the second year; and the refinement of the Draft Plan with a final version of the Plan in the third year, after a formal consultation phase. Work on additional topic reports continued in the second and third years of the commission. The annual Mersey Estuary Conferences served as milestones within the work programme, providing target dates for the completion of the main elements of the Plan, the opportunity for discussion and feedback at regular intervals and, over the three-year period as a whole, an evolving network of contacts and increased awareness of the range of issues connected with estuary management.

The Study Team took full advantage of the opportunities offered by its location in a graduate school of planning (the Department of Civic Design). Graduate students taking the Department's professionally accredited planning degree (the Master of Civic Design) were able to work as research assistants for periods of six to eight weeks at various stages throughout the commission and contributed in a major way to the documents that have been produced. Academic staff provided their time and professional skills without payment and this enabled the limited budget available for the Plan to be channelled to a large extent to the student research assistants. Nineteen graduate students benefited in this way during the commission.

Quite apart from the contributions that students made directly to the Plan, there were benefits for the teaching programme of the Department, in project work, field studies and in Masters degree dissertations.

Throughout the three-year commission, the Study Team worked closely with a Technical Steering Group, drawn from the membership of the Estuary Project Group. Steering Group members served in an individual capacity and provided detailed guidance on all of the technical aspects of the Plan, giving freely of their own time and readily sharing information and expertise.

The University Study Team, in carrying out its work, had to cooperate with an enormously wide range of organisations, many with diametrically-opposed views about the Estuary and its future development. This called for a non-partisan, sensitive and diplomatic approach. The Study Team quickly recognised that there are many specialists on particular topics connected with the Estuary, but very few generalists capable of taking an holistic view. It therefore sought to cultivate general knowledge and understanding about the Mersey and to promote dialogue between the various specialists. By the time of the Fourth Estuary Conference, in March 1995, there was clear evidence that this dialogue was starting to develop.

The University feels that it has benefited significantly from the experience of taking on such a substantial and wide-ranging commission. The Estuary Management Plan has provided a challenging opportunity for the University to play a useful role in the regional community, for individual staff to engage in Continuing Professional Development, and for future recruits to the planning profession to gain valuable practical experience on an innovative planning study.

Figure 9: Work programme 1992 -1995.

	March 1992	March 1993	March 1994	March 1995
Conference	First Report	Draft Plan	Final Plan	
Public Sector Partners		Report to Committee	Report to Committee	Approval of Plan
Private and Voluntary Sector		Consultation: Area Workshops	Consultation	Consultation: Area Workshops
University Study Team	• Definition of Study Area • Voluntary / Private Sector Consultation • Statutory Agency Review • Navigation, Tidal Regime and Level of Use • Nature Conservation and Water Quality • Recreation • Review of Experience Elsewhere	• Prepare Proposals • Advise on Mechanisms • Prepare Remaining Topic Reports	• Revise Plan	• Prepare Plan for Formal Publication
Public	Conference 1	Conference 2	Conference 3	Conference 4

APPENDIX 4

A Complete List of Documents Produced by the University Study Team

Topic Reports

	Number of pages
1 Initial Consultation with Voluntary Organisations and the Private Sector	44pp
2 Navigation, Tidal Regime and Level of Use	36pp
3 Water Quality and Nature Conservation	69pp
4 Informal Recreation Opportunities	47pp
5 Initial Consultation with Statutory Agencies	119pp
6 Landownership and Tenure	29pp
7 The EC Urban Waste Water Treatment Directive	73pp
8 Water-based Recreation	57pp
9 Tourism	119pp
10 Emergency Planning	40pp
11 Fishing and the Mersey Estuary	34pp
12 Coast and Flood Defence	59pp
13 The Implementation of an Estuary Management Plan: Organisational Structures and Institutional Arrangements	86pp
14 Monitoring the Management Plan	35pp
15 Estuary Projects: An Outline Programme	25pp

Area Issue Reports

1 The Upper Estuary	44pp
2 The Inner Estuary	59pp
3 The Outer Estuary	49pp

Overview Reports

	Number of pages
Mersey Estuary Management Plan: First Year Report	52pp
Mersey Estuary Management Plan : Second Year Report : Initial Proposals	69pp
Mersey Estuary Management Plan : Summary of the Draft Plan	24pp
Mersey Estuary Management Plan : Draft Plan for Consultation	87pp
Mersey Estuary Management Plan : Report on the Consultation Exercise	41pp
Mersey Estuary Management Plan : Third Year Report and Draft Plan	86pp

Obtaining Copies of the Documents

Copies of any of the above documents may be obtained by writing to:

Professor Peter Batey,
Mersey Estuary Management Plan,
Department of Civic Design,
University of Liverpool,
Liverpool L69 3BX.

A charge will be made for photocopying and for postage and packing.

MAP 4 THE KEY MAP

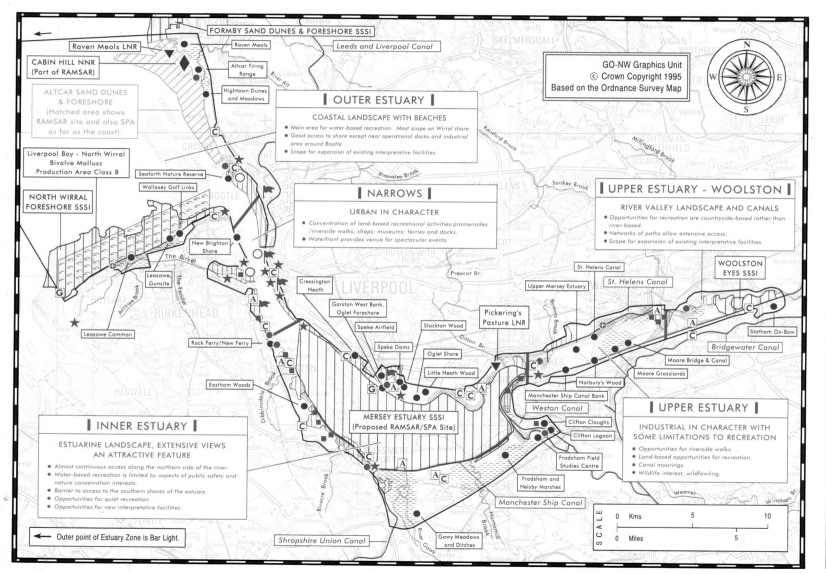